T0241998

GEOMETRIC SYMMETRY

TO HILDA AND ANNA

All ornament should be based upon a
geometrical construction.
Owen Jones: *The Grammar of Ornament*

Geometric
Symmetry

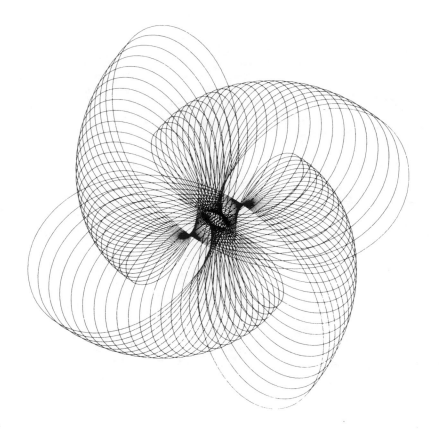

E.H. LOCKWOOD AND
R.H. MACMILLAN

CAMBRIDGE
UNIVERSITY PRESS
Cambridge
London - New York - Melbourne

CAMBRIDGE UNIVERSITY PRESS
Cambridge, New York, Melbourne, Madrid, Cape Town, Singapore, São Paulo, Delhi

Cambridge University Press
The Edinburgh Building, Cambridge CB2 8RU, UK

Published in the United States of America by Cambridge University Press, New York

www.cambridge.org
Information on this title: www.cambridge.org/9780521216852

© Cambridge University Press 1978

This publication is in copyright. Subject to statutory exception
and to the provisions of relevant collective licensing agreements,
no reproduction of any part may take place without the written
permission of Cambridge University Press.

First published 1978
This digitally printed version 2008

A catalogue record for this publication is available from the British Library

Library of Congress Cataloguing in Publication data

Lockwood, Edward Harrington.
Geometric symmetry.

Bibliography: p.
Includes index.
1. Geometry. 2. Symmetry. 3. Symmetry groups.
I. Macmillan, Robert Hugh, joint author. II. Title.
QA447.L63   516′.1   77-77713

ISBN 978-0-521-21685-2 hardback
ISBN 978-0-521-09301-9 paperback

Contents

ACKNOWLEDGEMENTS

Thanks are due to the following for permission to reproduce illustrations.

Michael Franses Fig. 0.03

The Victoria and Albert Museum Fig. 0.04

Escher Foundation, Haags Gemeentemuseum, The Hague Figs. 0.05, 11.14, 11.15

Etienne Revault, Photographer Fig. 6.11

The Mathematical Association Figs. 12.14, 13.02

Mr J.D. Connolly Fig. 12.15

The 'Altair Design', illustrated on Fig. 13.06, was developed by Dr Ensor Holiday and published by Longman in 1973

The computer graphics for the designs on the cover and preliminary pages were kindly provided by Dr A.C. Norman

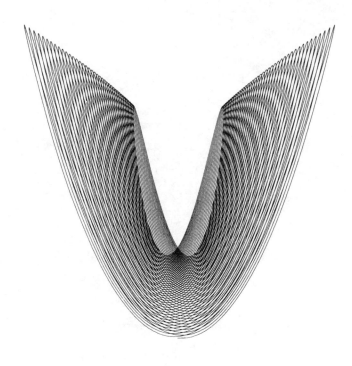

Preface

Symmetry is of interest in two ways, artistic and mathematical. It underlies much scientific thought, playing an important role in chemistry and atomic physics, and a dominant one in crystallography. It is important in architectural and engineering design and particularly in the decorative arts. Yet the literature available is comparatively sparse. Mathematically it has been extensively treated by continental authors, particularly, in recent times, in Russia. Books on crystallography naturally devote considerable attention to it, giving at least a descriptive account, and the *International Tables for X-ray Crystallography* (vol. 1) is an invaluable work of reference. For the general reader there is Weyl's *Symmetry*, a delightful and stimulating book, and more recently Rosen's *Symmetry Discovered*: but beyond these there is little more than a few short sections in mathematical books that deal mainly with other topics.

In this book we attempt to provide a fairly comprehensive account of symmetry in a form acceptable to readers without much mathematical knowledge or experience who nevertheless want to understand the basic principles of the subject. It is hoped that it will be found useful in school and other libraries and as preliminary reading for students of crystallography. The treatment is geometrical, which should appeal to art students and to readers whose mathematical interests are that way inclined. It is also hoped that the full enumeration of symmetry types will make it useful for reference purposes.

Part I is largely descriptive and is written with the non-mathematical reader in mind. It gives a general account of the subject and indicates how the ideas may be applied to the construction of new designs from very simple elements. Part II is more mathematical, but only the most elementary knowledge of geometry is assumed. The more systematic treatment in this part makes it possible to enumerate and classify the symmetry groups of each kind, using throughout the convenient notation that has been evolved by crystallographers during the last 50 years and is now accepted internationally.

The results in this book are not new, though some of them have been available only in a form not readily accessible. The approach adopted, however, stems from a basic reappraisal of the subject from the geometrical point of view. Our intention, in dividing the book into two parts, has been firstly to reveal the simplicity and beauty of the underlying ideas and then to show that the geometrical approach can lead to a consistent mathematical development of the subject.

Our thanks are due to Dr A.G. Howson, who read the first draft, and to the Press for their ready and patient cooperation and for tackling so successfully the problems that arose in producing the book.

E.H.L.

R.H.M.

November 1977

Historical note

In 1611 Kepler wrote a little monograph *Strena seu de Nive Sexangula* (A New Year Gift: On Hexagonal Snow) in the course of which he considered the packing together of circles in a plane and spheres in space. This was printed, but attracted little attention. The idea that external form might depend on internal structure appeared again, however, as soon as crystals began to be seriously studied. Hooke (*Micrographia*, 1665) thought they might be built up of spheroids. Bartholinus (1669) studied calcite (Iceland spar), noticing the double refraction produced by the crystals and the rhomboidal cleavage planes. He measured the angles between the faces. Nicolaus Steno studied the geometrical forms of crystals and noted the constancy of the angles. Huygens (*Traité de la Lumière*, 1690) tried to explain the calcite phenomena by supposing the crystals to be built up of closely packed spheroids. Bergman (1773) and Haüy (1782) thought in terms of a structure of brick-like elements (*molécules soustractive*, as they were called), but these ideas were slow to develop because of the lack of an adequate atomic theory. Seeber (1824) supposed molecules to be spaced out at intervals, but he was ahead of his time and only in 1879 was the idea revived, by Sohncke. Meantime Hessel (1830) and Axel (1867) had classified crystals into 32 classes and Bravais (1850) had described the 14 space lattices known by his name. In 1890 the 230 space groups were enumerated by Fedorov, and independently by Schoenflies.

The results for line groups and plane groups were implicit in this, or at least could be easily derived. But the interest in symmetry was so entirely concentrated on its application to crystallography that such simple matters as the 7 'frieze' groups and the 17 'wallpaper' groups were not specifically dealt with until much later, notably by Polya[1] and Niggli[1] in 1924 and by Speiser[2] in 1927. Speiser also enumerated the 75 crystallographic line groups in three dimensions and the 31 'ribbon' groups. The 80 'layer' groups were dealt with two years later by Alexander and Herrmann,[3] and also by Weber.[3] In 1930 Heesch[4] and Shubnikov[4] both wrote on the symmetry of continuous and semi-continuous figures.

Mathematically it might have seemed that this was the end of the road, but there was to be a further development. In 1930 Heesch[4] suggested the possibility of four-dimensional symmetry groups in three-dimensional space. The extra dimension could be represented by a change of colour. H.J. Woods[5] (1935) discussed the same possibility under the name 'Counterchange symmetry' and listed the 17 types of particoloured friezes ('counterchange borders'). The idea was taken up by Shubnikov in 1945 and developed by him in *Symmetry and Antisymmetry of Finite Figures* (1951). This dealt with point groups only. Cochran[6] in 1952 listed the 46 particoloured plane groups in two dimensions, deriving them from the 80 layer groups enumerated earlier by

Alexander and Hermann. In 1953 Zamorzaev enumerated the 1651 dichromatic space groups, calling them the *Shubnikov groups*. These were obtained independently in 1955 by Belov, with Nerovina and Smirnova.

As the term *antisymmetry* implies, Shubnikov's extra dimension consisted of a polarity and could be represented by two colours. The idea was later extended to three or more colours by Belov and others.

The study of symmetry, like that of some other branches of mathematics, has been confused by a multiplicity of notations. Fortunately the International Union of Crystallography has established a standard notation, now widely accepted, for the uncoloured groups. For dichromatic change a simple extension, first used by H.J. Woods,[5] has been generally adopted, but there is as yet no recognized notation for polychromatic groups.

References:
1 *Zeitschrift für Kristallografia*, **60** (1924)
2 Speiser: *Die Theorie der Gruppen* (1927)
3 *Zeit.f.Krist.* **70** (1929)
4 *Zeit.f.Krist.* **73** (1930)
5 *Journal of Textile Inst.* **26** (1935)
6 *Acta Crystallagr.* **5** (1952)

Introduction

The idea of symmetry is familiar in everyday life, whether applied to solid objects or to patterns and designs. The essential feature of a symmetrical object is that it can be divided into two or more identical parts: and furthermore that these parts are systematically disposed in relation to one another. In addition some objects, such as a ladder or a wallpaper pattern, have repetitive elements, whereas others, for example a table or a handcart, though clearly symmetrical, are not repetitive.

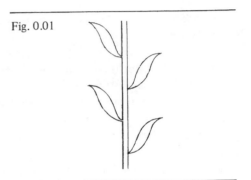

Fig. 0.01

Fig. 0.02

One might well ask whether there can exist an unlimited variety of symmetrical patterns and objects or whether it is possible to classify them, grouping them according to the characteristics they have in common. The aim of this book is to show how such a classification can be undertaken, and indeed to go a step further by enumerating all the possible types of symmetry, both for patterns and solid objects, either with or without repetitive elements.

We shall find that, while a single object may exhibit any one of an infinite range of symmetry types, there are severe limitations on the number of types of repetitive pattern. Thus it will appear that there are only 7 types of 'frieze pattern', 17 of 'wallpaper pattern', and so on. Many varieties of symmetry are found in decorative art and in the natural and other forms that surround us. But some natural forms, such as those of shells and plants, exhibit a structure in which parts are similar but differ in size, often diminishing at a uniform rate. Such forms, which are characteristic of many growth patterns, are mentioned briefly in Chapter 10.

The two main kinds of symmetry found in plane patterns are based on reflexion and rotation. That is, one part of the pattern can be brought into coincidence with another part by one of these two means. Examples of symmetry by reflexion are the letter A and the Latin cross: patterns having rotational symmetry are the letter Z and the swastika. It is also possible for a pattern to exhibit both kinds of symmetry together, as in the case of the Greek cross or the letter I. With repetitive plane patterns, such as that of a parquet floor, we shall find that there is a third kind, called *glide reflexion* symmetry (Fig. 0.01).

Solid objects too can have reflexional and rotational symmetry and will often have both. We shall show that in addition there are several other kinds of symmetry in three dimensions. First there is *inversion* symmetry, as exhibited by the pair of cranks of a bicycle; such symmetry may be found in conjunction with one or both of the other types. Next there is *rotatory inversion*. A simple example of this is provided by two equal sticks laid across each other at right angles (Fig. 0.02). If the cross were turned through a right angle and then inverted, the sticks would have changed places. Finally, repetitive three-dimensional patterns may have *screw symmetry*, an example of which is seen in a 'spiral' staircase. As a result of the existence of all these possible symmetries it is not surprising that the symmetry types for three dimensions are much more numerous than those for two.

The physical process of growth, be it of crystals or of living matter, favours the production of symmetrical forms; and indeed the world is pervaded by shapes having a greater or lesser degree of symmetry, from the leaf to the human form. For this reason there must be a sense of symmetry in the mind of every artist. In the applied arts, from the simplest hand-made pot to the Palladian country house or the Gothic cathedral, symmetry appears very often as a prominent feature of the design; and when it does not, there has usually been a deliberate and conscious effort to avoid it. In the design shown in Fig. 0.03 the carefully contrived avoidance of symmetry (in the pattern taken as a whole) is so obvious as to make one realise that symmetry and repetition come naturally: it is the avoidance of them that is artificial.

It is in Arabic and Moorish design that the different kinds of symmetry have been most fully explored. The later Islamic artists were forbidden by their religion to represent human or even animal form, so they turned naturally to geometrical elaboration. Among their works, and also among those of ancient Egypt, numerous examples are to be found of all the 7 types of 'frieze' pattern and most, if not all, of the 17 possible types of 'wallpaper' pattern.

Fig. 0.03

Friezes and wallpapers are examples of two-dimensional patterns that are repeated indefinitely in one, or two, directions respectively. Their possible types can be described mathematically in terms of two-dimensional *line groups* and *plane groups*. These are groups of movements that bring the whole pattern into self-coincidence. There are also two-dimensional patterns that are not repeated and they are described by means of *point groups*, that is, groups of movements that leave one point fixed. Patterns having various types of point symmetry include letters of the alphabet, many commercial symbols, the patterns produced by the kaleidoscope, and the rose windows of cathedrals. As well as in friezes, two-dimensional line symmetry is found in the decorative borders of oriental rugs and in the edges of mosaic pavements. Similarly, two-dimensional plane groups describe the decorative patterns applied to surfaces, as seen in wallpapers, tessellated pavements, mosaics and tiled arabesques, and the repetitive patterns of many carpets and rugs. Needlework, too, offers many examples. Fig. 0.04 shows a mid-nineteenth-century English sampler in which many types of frieze and wallpaper pattern can be seen.

Three-dimensional point groups can be used to classify the many types of solid object having point symmetry. Lampshades and chandeliers can have very interesting symmetry properties, as also have many items of jewellery, such as bracelets and brooches. Of natural forms, the 32 classes into which crystals are grouped according to their external shapes correspond to the 32 'crystallographic' point groups. Among the most familiar of these are the cubic crystals of common salt and the six-pointed crystals of snow.

Three-dimensional line symmetry is that possessed by objects having repetition in one direction only, such as ropes, chains, necklaces and plaits. Line symmetry is also found in carved or decorated columns and straight staircases and balustrades, and in the structure of some buildings, particularly in arcades and cloisters. It is a feature too of many important chemical molecules such as those of proteins and polymers and of DNA, the basis of life. The growth pattern of many plants shows approximate line symmetry along the stem.

Fig. 0.04

Plane symmetry in three dimensions is the property of almost all fabrics, including woven and knitted materials, also of carpets and rugs, basket work and cane work. In the realm of architecture we find that the stone screens of Moslem buildings, as well as brick walls and roof tiling, exhibit various types of plane symmetry.

Finally, in three dimensions, it is possible to have patterns repeated indefinitely in all three directions, or throughout space. Such patterns are described by *space groups*, of which there are no less than 230. Few of them are met with in ordinary experience; however, they are fundamental to matter in the solid, since any crystalline material has an internal structure based on one of these groups. (It has been shown during the present century, by means of X-rays, that the atoms in a crystal are always arranged in a regular formation belonging to one of the 230 groups. If they were not so arranged the material would be not crystalline but amorphous, like glass.) The nearest approach to space symmetry in everyday life is in the various methods available for packing similar objects, such as tennis balls, into a box, or of piling bricks into a stack. The mathematically minded can also consider the symmetry of space tessellations, that is, the various ways of filling space with one or more kinds of identical solids.

In the free arts of painting and sculpture perfect symmetry is more often avoided than pursued, being replaced usually by a sense of balance, an equality of 'weight' rather than of measurement. However, some modern artists, notably M.C. Escher and M. Vasarely, have used various degrees of plane and point symmetry to very fine effect. Fig. 0.05 shows one of M.C. Escher's designs and two more are illustrated in Figs. 11.14 and 11.15.

Fig. 0.05

One of the fascinations of the study of symmetry lies in the relationships revealed between a wide variety of objects, natural and artistic. A snow crystal, the bee's cell, a patchwork quilt and the basalt rocks of the Giant's Causeway have something in common, as represented by the word 'hexagonal', yet their forms differ in certain important respects. The mathematical concept of symmetry provides a classification for such similarities and differences and can thus be used as a common basis of description for much of the visual world.

Part I Descriptive

To describe the symmetry of a pattern or object one must consider its division into identical, or congruent, parts and the way in which those parts are related positionally to one another. It is natural and convenient to describe such relationships in terms of movements. Arrows pointing north, south, east and west suggest the movements of a weathercock, and the arms of the Isle of Man suggest rotations through 120°. A repeating pattern, such as a frieze, suggests translatory movements, the movements of a procession.

All these are physical movements, but a pair of wrought iron gates suggests a reflexion, which is not exactly a movement, unless we count Alice's movement through the looking-glass. We shall, however, find it convenient to give the word *movement* this slightly extended meaning. A left hand and a right hand can be placed so that each is a reflected image of the other and we shall call the change from one to the other a 'movement', although it is not possible for the right hand to be moved physically into the space the left hand was occupying.

These movements, and some combinations of them, enable us to describe in the following chapters the various kinds of symmetry. But the movements also serve another purpose. They are the means by which complicated symmetrical patterns can be built up from simple elements. This is the principle of the kaleidoscope, which uses two reflexions to produce patterns containing rotations and further reflexions as well. Thus two movements in combination imply others.

Such relationships between movements will form a main part of our study. They determine what is at choice and what follows automatically. This is a question of interest to the designer and it will receive some attention in Part I. It is also the kernel of the mathematical approach, as will be apparent to the reader of Part II.

1

Reflexions and rotations

The front view of the human face, as it appears in a drawing, is a good, if not perfect, example of *symmetry about a line*. Every point A on one side of the line has its counterpart A' on the other side (Fig. 1.01) such that AA' is bisected at right angles by the line. It is natural to think of the line as a mirror and of A and A' as mirror images of each other. We shall call the line a *mirror line* or *reflexion line* and we shall use the word reflexion to mean the change from one side of the figure to the other.

This is in two dimensions: but the human face is really three-dimensional and a mirror is really a plane, so in three dimensions we have a *mirror plane* rather than a mirror line. The 'movement' which we call *reflexion* changes A to A' and A' to A, thus leaving the whole figure self-coincident or, as we may conveniently say, unchanged.

The two sides of the figure are *congruent*. This word means that two figures, or parts of a figure, are so related that for every point of one there is a corresponding point of the other and that the distance between any two points of the one is equal to that between the corresponding points of the other. It follows that corresponding angles are equal, but it is to be noted that, in the case of reflexion, the turns represented by corresponding angles are in opposite senses, clockwise and anticlockwise (Fig. 1.02); and in three dimensions corresponding turns are right-handed and left-handed (Fig. 1.03). The two parts of the figure are said to be *indirectly* or *oppositely congruent*.

Rotation suggests another kind of symmetry (Fig. 1.04) and here the congruence is *direct*, each part of the figure being an exact reproduction of another part, though differently placed. Rotation is a *proper* or *direct movement*. Here again it should be noted that the turn, applied to the figure as a whole, leaves it unchanged.

A movement or transformation that changes a figure into a congruent figure is called an *isometry*. Rotation is a *direct isometry* and reflexion is an *indirect* or *opposite isometry*. *Symmetry* means that the parts of a figure are not only congruent* but related by an isometry, e.g. reflexion or rotation, in such a way that the whole figure is self-coincident under that isometry. A *symmetry movement* is one that changes every part of a figure into another part, leaving the figure as a whole unchanged. Obviously it may be repeated any number of times with the same effect.

The circle is the most perfectly symmetrical plane figure, because it can be turned about its centre through any angle whatever or reflected in any diameter. Similarly in three dimensions a sphere can be turned through any angle about any diameter or reflected in any plane through the centre. Apart from figures consisting entirely of concentric circles or spheres there is always a smallest

*Here we exclude the somewhat exceptional case of similarity. See Chapter 10.

Fig. 1.01

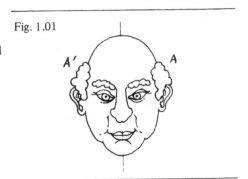

Fig. 1.02

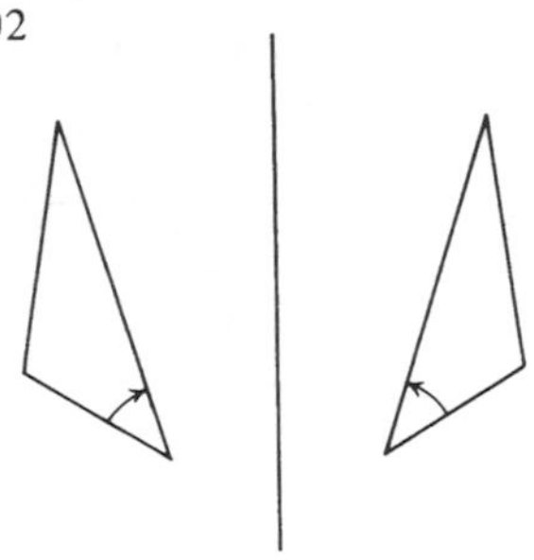

Fig. 1.03

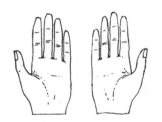

Fig. 1.04

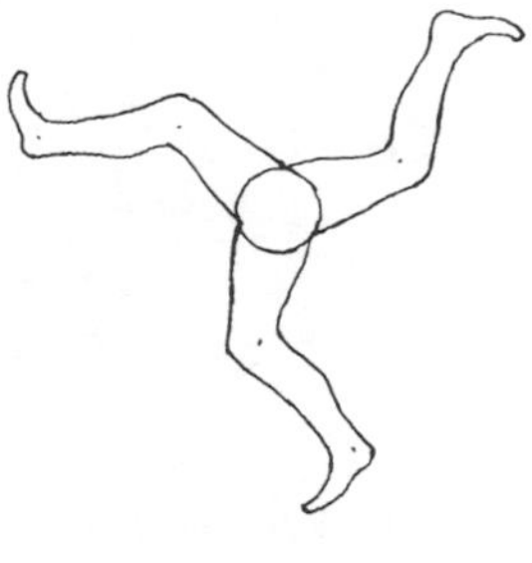

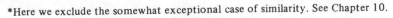

possible angle of rotation. The element of pattern can be rotated into 2, 3, 4 or more positions, by half-turns, one-third-turns, quarter-turns, etc. We call these rotations *diad*, *triad*, *tetrad*, and so on, accordingly.

Fig. 1.05

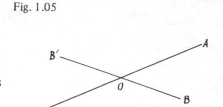

In two dimensions rotation is about a point; in three dimensions about an axis. In either case the isometry is direct. There is another movement, however, called *central inversion* (or simply *inversion*, for short), which in two dimensions is a direct isometry and in three is an opposite one. Inversion is reflexion in a point. Inversion in a point O (called the *centre of inversion*) consists of the replacement of any point A by its image in O, i.e. AO is produced to A' so that $AO = OA'$ (Fig. 1.05). Similarly B is replaced by its image B'.* It is easily seen that in two dimensions inversion is the same as a half-turn, as, for example, in the silhouette pattern of two hands shown in Fig. 1.06(*a*). But if the hands are three-dimensional, as in Fig. 1.06(*b*), a right-handed turn of one corresponds to a left-handed turn of the other, showing that inversion in three dimensions is an opposite isometry. It is in fact a combination of rotation and reflexion: a half-turn about an axis and a reflexion in a plane perpendicular to that axis. This can be seen in Fig. 1.07, where A is rotated to A' and then reflected to A''. It can also be demonstrated with two hands.

Fig. 1.06

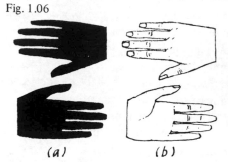

(*a*)   (*b*)

This shows that it is necessary to consider not only reflexions and rotations but also the various ways in which they can be combined. Moreover, for repeating patterns the basic movements include translations and these will combine with reflexions and rotations to form further movements such as glide reflexions and screw rotations. These we shall consider in due course, dealing first with movements in two dimensions.

Fig. 1.07

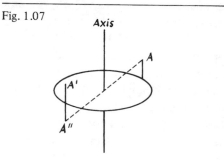

---

*This transformation is not to be confused with inversion in a circle or a sphere. Algebraically it is $(x, y, z) \rightarrow (-x, -y, -z)$, whereas inversion in a sphere is $(x, y, z) \rightarrow (k^2/x, k^2/y, k^2/z)$.

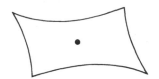

Fig. 2.01

2

Finite patterns in the plane

Fig. 2.02

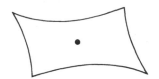

Fig. 2.03

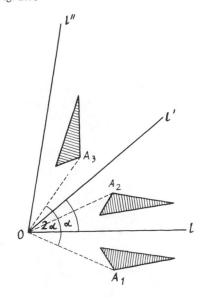

Looking at finite symmetrical patterns such as those of Figs. 2.01, 2.02 and 2.03, we notice that there is always a centre point or a centre line. In the symmetry movements of these patterns, i.e. in the rotations and reflexions that leave them unchanged, the centre point (or line) remains fixed. The symmetry of such patterns is called *point symmetry*.

By contrast a repeating pattern, as in a frieze or a wallpaper, is built up by translations, which move every point the same distance and in the same direction. The pattern is then in theory infinite.

The only movements that leave a point fixed are rotations about the point and reflexions in lines through it. We consider rotations first and we suppose that there is a minimum angle of rotation. (This excludes designs made up entirely of circles centred on the fixed point.) As a symmetry movement leaves the pattern unchanged it can be repeated any number of times. Thus a turn through 90° means that one of 180° or 270° is equally possible. Hence the minimum angle of rotation must be a sub-multiple of 360°, say $360°/n$, where n is an integer. (For proof of this see p. 106.) Patterns of this sort are said to have *cyclic symmetry*, and there are an infinite number of types, according to the value of n. Figs. 1.04, 2.01, 2.02 show examples in which $n = 3, 4, 2$. These types are sometimes denoted by the symbols $C_3, C_4, C_2, \ldots$, but we shall more often use simply the numbers $3, 4, 2, \ldots$, according to the agreed 'international' notation for symmetry groups.*

Fig. 2.04

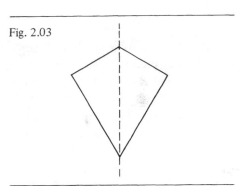

Next we consider reflexions. There may be a single reflexion line, as in Fig. 2.03, and this is perhaps the simplest of all symmetry types. If, however, there is more than one, there must also be a rotation centre. Note first that, for a finite pattern, there cannot be parallel reflexion axes, as that would imply a translation (Fig. 2.04) and hence a repeating pattern. If then the mirror lines intersect at angle a it can be seen from Fig. 2.05 that the two reflexions combine to give a rotation of amount $2a$ about the point of intersection. (A_1 is reflected in l to A_2 and thence in l' to A_3.)

Fig. 2.05

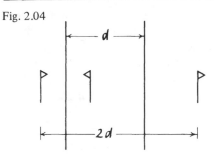

It is equally true, and shown in the same figure, that a rotation about a point and a reflexion in a line through that point combine to give a reflexion in another line through the point. (If A_3 is rotated about O to A_1 and then reflected in l to A_2, the combined movement is a reflexion in l'.) Moreover other reflexion lines are implied, at angular intervals $a, 2a, \ldots$, from l and l'. (If A_3 is rotated as before to A_1 and then reflected in l', the combined movement is reflexion in a line l'' at angle a to l'. Thus the mirror line l is itself reflected to a new mirror line l''.)

*We shall refer to the symbols C_3, C_4, etc. as being in the *group notation*. See also the Index of Groups.

Fig. 2.07

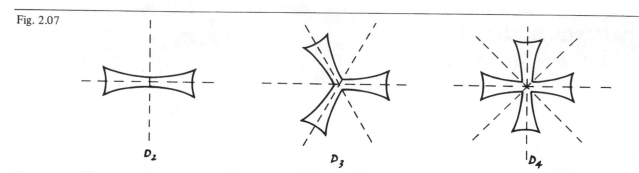

D_2 D_3 D_4

So, if a pattern admits reflexions in two lines intersecting at O at an angle a, it will also admit rotation about O through $2a$ and reflexions in further lines through O at intervals a. In short, the reflexion lines rotate with the figure, and are reflected in one another (Fig. 2.06).

Fig. 2.06

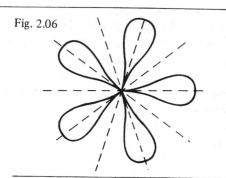

Patterns of this kind are called *dihedral*. The simplest of these types is that illustrated in Fig. 2.03, with one mirror line. The group symbol for this is D_1, while that of Fig. 2.06, with its pentad rotation and five mirror lines is D_5. Clearly there are an infinite number of dihedral symmetry types. We illustrate D_2, D_3 and D_4 in Fig. 2.07.

The 'international' symbols for these symmetry types give first a number to indicate the rotation, followed by letters m for the mirror lines. It will be noticed that in D_4 the four mirror lines divide into two sets, those along the arms of the cross and those bisecting the angles so formed. Each set is related by the tetrad rotation. This happens with all the even-numbered types, but with D_3, and all odd-numbered types, the two sets coincide. For this reason D_4 is given the symbol $4mm$, but D_3 is $3m$. The two systems may be compared as follows:

$$D_1 \quad D_2 \quad D_3 \quad D_4 \quad D_5 \quad D_6 \quad . \quad . \quad .$$
$$1m \quad 2mm \quad 3m \quad 4mm \quad 5m \quad 6mm \quad . \quad . \quad .$$

There are thus two infinite sets of symmetry types for finite patterns in two dimensions, the cyclic types and the dihedral types.

3

Frieze patterns

A frieze pattern, as, for example, that shown in Fig. 3.01, is one consisting of a *motif*, or element of pattern, repeated at regular intervals along a line, *ad infinitum*. In practice, friezes are not infinite in length, but our theory is concerned with movements that can be repeated any number of times, so it is convenient to suppose that patterns are extended indefinitely. (When we give examples of repeating patterns we shall always suppose them to be extended to infinity.)

The basic example of a frieze pattern is an infinite row of dots (Fig. 3.02). Any other frieze pattern can be built up by replacing the dots of such a *row* by a motif, or repeated pattern element. The row of dots represents the mode of repetition. Note the translatory movement that takes an element of pattern from one position to the next. If this movement were applied to the pattern as a whole it would leave it unchanged, but only assuming that the pattern extends indefinitely.

For the frieze to have rotational or reflexional symmetry the motif must have that symmetry and so must the row of dots. In Fig. 3.03 the motif has rotational symmetry 6, but the frieze as a whole would not remain unchanged if turned through 60°, so we cannot say that the frieze has hexad symmetry. (It does have diad symmetry.)

Fig. 3.01

Fig. 3.02

• • • • • • • • • • • • •

Fig. 3.03

* * * * * * * * * * * * *

The row of dots could be considered as one dimensional and it then has the only kind of one-dimensional symmetry, namely reflexion in a point, which we call *inversion*, the point being the *centre of inversion*.* For the row of dots the centre of inversion can be any dot or the mid-point of the interval between two dots. There are just two types of one-dimensional frieze: those that have this symmetry (Fig. 3.04(*a*)) and those that do not (Fig. 3.04(*b*)). More often a frieze is made with a two-dimensional pattern element, as in Fig. 3.01, and we must then regard the row as set in two-dimensional space. The row of dots then has several kinds of symmetry, namely:

inversion (which we now regard as rotational symmetry by half-turns about any of the points mentioned),

transverse reflexion (i.e. reflexion in lines perpendicular to the row, drawn through the same points),

longitudinal reflexion (i.e. the line of the row may itself be regarded as a mirror line).

The motif, and hence the frieze as a whole, may have any or all of these symmetries. This gives five types of frieze, as shown in Fig. 3.05. In the symbols the prefix *r* indicates repetition along a row;** the numeral in the first position

*See footnote p. 10.

**In the 'International' notation p is used, rather than *r*.

Fig. 3.04

(*a*)

(*b*)

Fig. 3.05

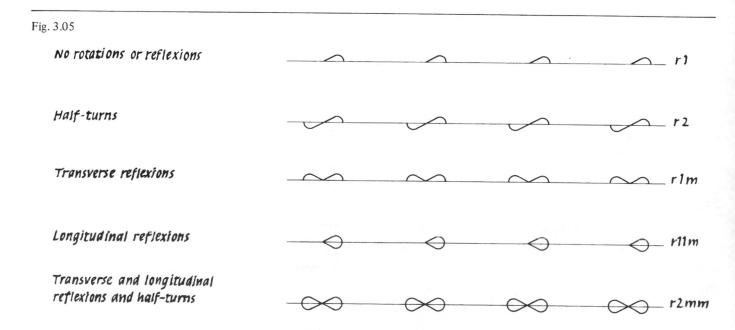

No rotations or reflexions — r1

Half-turns — r2

Transverse reflexions — r1m

Longitudinal reflexions — r11m

Transverse and longitudinal reflexions and half-turns — r2mm

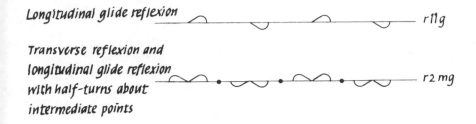

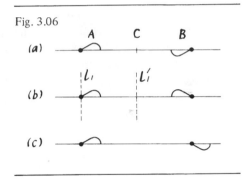

Fig. 3.07

Longitudinal glide reflexion ———— *r11g*

Transverse reflexion and
longitudinal glide reflexion ———— *r2mg*
with half-turns about
intermediate points

after the prefix refers to rotation (2 meaning that half-turns are admissible, 1 that they are not); *m* in the second position indicates transverse mirror lines (or a 1 is used if there are none); while *m* in the third position shows a longitudinal mirror line.

It will be noticed that we may have one of these symmetries or all three. Any two imply the third. This is because of the theorem illustrated in Fig. 2.05 (but with a equal to 90°), which shows that reflexions in lines at right angles combine to give a half-turn; and that a half-turn combined with either reflexion gives the other. For this reason the symbol *r2mm* is sometimes shortened to *rmm* (it could just as well be *r2m*).

We must also consider the combination of the translation with any of these movements. If a pattern is given a half-turn about a point A and then moved forward a distance AB (Fig. 3.06(*a*)) the combined movement is equivalent to a half-turn about an intermediate point C; and similarly a transverse reflexion in a line l (Fig. 3.06(*b*)), combines with the same forward movement to give a reflexion in a parallel line l'. But something new emerges when we consider a translation combined with a longitudinal reflexion (Fig. 3.06(*c*)). This movement is called a *glide reflexion* (symbol *g*) and it gives rise to two further types of frieze pattern (Fig. 3.07). These may be regarded as formed from the last two of the five earlier types by splitting the pattern into two and shifting one part forward half a unit (i.e. half the repetition distance).

This makes altogether seven types of two-dimensional frieze pattern, and exhausts the possibilities. Examples of these types are shown in Fig. 3.08.

Fig. 3.06

Fig. 3.08

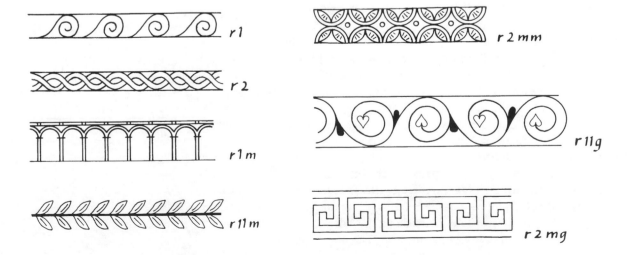

r1

r2

r1m

r11m

r2mm

r11g

r2mg

4

Wallpaper patterns

Fig. 4.01

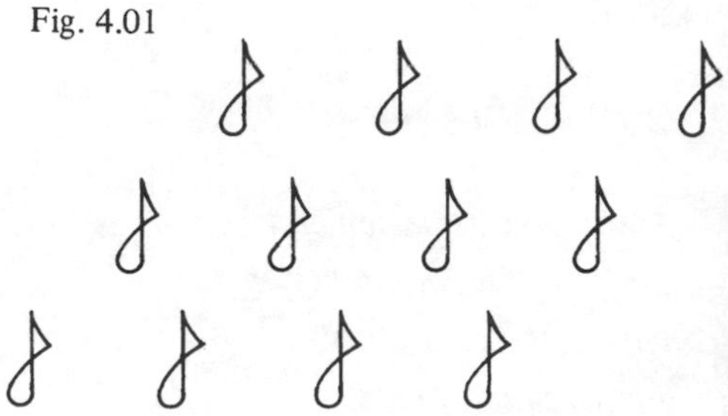

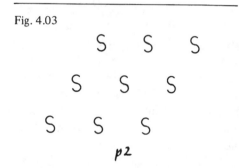

Fig. 4.02

A *wallpaper pattern* is one consisting of a motif repeated at regular intervals in more than one direction (Fig. 4.01). The simplest example is formed by a row of dots repeated in parallel rows, as in Fig. 4.02. This is called a *net*. But the net is not only the simplest example. It is the framework on which any wallpaper pattern can be built. It represents the mode of repetition. Any motif can be used, provided it is repeated in the way indicated by the net.

Fig. 4.03

The repetition is in theory endless, the pattern extending to infinity in all directions. Our illustrations are necessarily finite, but it must be supposed that they are extended indefinitely.

The net may be thought of as built up from a single dot A and two translations T_1, T_2, as shown in the figure. For a given net, T_1 and T_2 can be chosen in a variety of ways. Applied to the dot A they suggest two sides of a parallelogram, the fourth vertex being reached from A by the combination $T_1 T_2$ (or $T_2 T_1$). This parallelogram may be regarded as a unit cell of the net.

Fig. 4.04

If the translations, or any combination of them, are applied to the whole (infinite) net, it is unchanged. We may in fact redefine a wallpaper pattern as a pattern that remains unchanged under two translations not in the same direction.

The parallelogram that forms a unit cell may sometimes be a special one such as a rectangle, square or rhombus. Taking the general case first, the only symmetry of the parallelogram cell is inversion. This is the same (in two dimensions) as rotational symmetry 2, that is to say that a half-turn about the centre leaves the figure unaltered. The net as a whole has this symmetry, not only about the centres of the cells but also about the vertices and the mid-points of the sides. A wallpaper pattern based on this net may or may not have these symmetries (Figs. 4.03, 4.04 respectively). These are the first two kinds of wallpaper pattern. Among commercial wallpapers the second is by far the more common, the first being rarely seen. The two types are referred to as $p2$ and $p1$ respectively, the numeral representing the degree of rotational symmetry and the prefix p the repetition in two directions.

Fig. 4.05

The motif, or element of pattern, associated with each point of the net could of course have other symmetries, but if these are not symmetries of the pattern as a whole they will not count for our purposes. In Fig. 4.05, for example, the motif has reflexional symmetry but the pattern as a whole has not. The symmetries of the whole pattern are in fact limited by those of the net.

Fig. 4.06

With a rectangular net, i.e. one whose unit cell is rectangular (Fig. 4.06), reflexions are possible, the mirror lines being along the sides of the rectangles and also along parallel lines midway between them. The motif applied to this

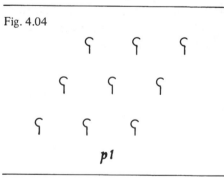

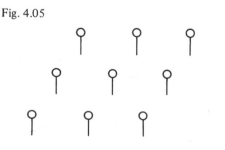

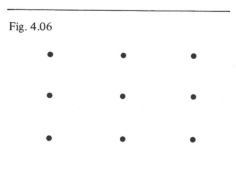

Fig. 4.07   Fig. 4.08

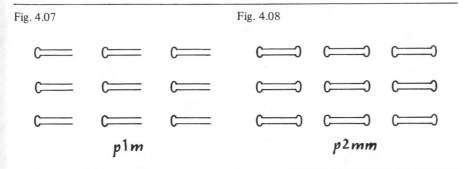

*p*1*m*   *p*2*mm*

Fig. 4.09

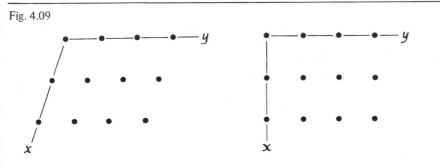

Fig. 4.10

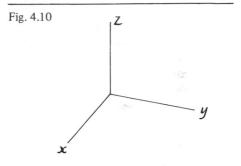

net may have reflexional symmetry with mirror lines in one or both directions (Figs. 4.07, 4.08 respectively).

It is not at this stage necessary to distinguish between the two directions, but as we shall need to do so later it is convenient to introduce coordinate axes here. They will not always be rectangular: it is better to have them in the directions of T_1 and T_2. In the diagrams we shall always draw the y-axis across the page, with the x-axis downwards (Fig. 4.09). This somewhat unusual arrangement is designed to fit in with diagrams in three dimensions with the x-axis coming towards the reader (Fig. 4.10).

In the symbols for the symmetry types (as used so far) the numeral in the first position after the prefix is for rotation about the origin and other points of the net. We now use m in the second position to represent reflexion in mirror lines perpendicular to the x-axis; and in the third position for mirror lines in some other direction. With a rectangular net this other direction is perpendicular to the y-axis. (The system is fully explained at the end of this chapter.)

It will again be noticed that, when there are reflexion lines in directions at right angles to one another, the half-turn is automatically admissible (Fig. 4.08) and for this reason the symbol $p2mm$ is sometimes shortened to pmm (it could just as well be $p2m$).

A somewhat similar connection between three movements is illustrated in Fig. 4.11. Here the two reflexions are in parallel mirror lines l_1, l_2. The combined movement, from A_1 to A_2 and thence to A_3, is equivalent to a translation of double the distance between the lines. Alternatively the translation, from A_1 to A_3, followed by reflexion in l_2, is equivalent to reflexion in l_1. This is the reason why, in Fig. 4.12, for example, the reflexion lines l, spaced at repetition distance, imply a further set l' halfway between them. It is in fact a regular feature of repeating patterns that a set of parallel reflexion lines (or planes) at repetition distance imply an intermediate set bisecting the intervals

Fig. 4.11

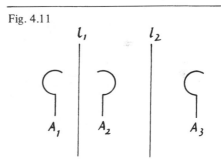

Fig. 4.12

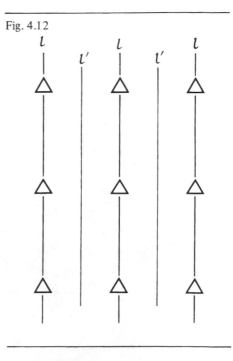

between them. In somewhat the same way, a set of intersecting reflexion lines, if they intersect at angles of rotational symmetry, imply an intermediate set bisecting the angles between them (Fig. 4.13). This follows from the theorem on p. 11.

As with frieze patterns, there are glide reflexion types related to those patterns in which a reflexion line runs in the direction of a translation. Thus the pattern of Fig. 4.07 suggests that of Fig. 4.14, the element of pattern being split in two. Here the intermediate lines also are lines of glide reflexion.

From the *2mm* type (Fig. 4.08) we derive two new ones, according as one or both reflexions are replaced by glides. These are shown (one cell only) in Fig. 4.15. The diad rotation is still possible, but the centres of rotation no longer lie at the intersections of the reflexion and glide reflexion lines. The complete systems of symmetries for these types are shown in Fig. 4.16, where the continuous lines are reflexion lines, the broken lines glide lines, and the lozenge symbols are centres of diad rotation. (If the glide lines in (*a*) were perpendicular to the *y*-axis, the symbol would be *p2mg*.)

Another kind of special parallelogram is the rhombus. Patterns based on a rhomboidal net (Fig. 4.17) are frequently found in wallpapers and other forms of decoration. The figure shows that the net may also be regarded as made up

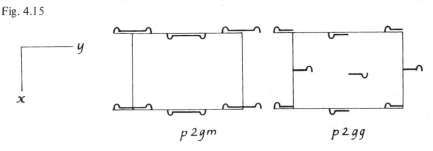

Fig. 4.13   Fig. 4.14

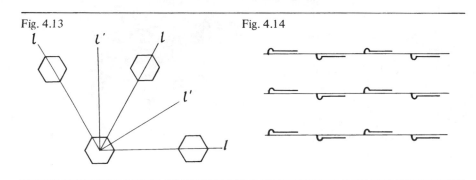

Fig. 4.15

p 2gm   *p 2gg*

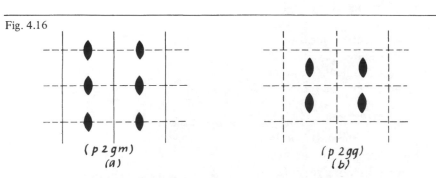

Fig. 4.16

(*p 2gm*)   (*p 2gg*)
(*a*)   (*b*)

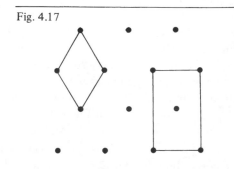

Fig. 4.17

of centred rectangles. This view we adopt, as it is more convenient when it comes to three-dimensional patterns. To distinguish the *centred* rectangles (Fig. 4.17) from the *primitive* ones (Fig. 4.06) we use the letters *c* and *p* respectively as prefixes to the symbols. Thus the types shown in Figs. 4.07, 4.08 are called *p1m* and *p2mm*, but those of Fig. 4.18 are *c11m* and *c2mm*.

With the centred rectangular net it is again possible to have reflexion lines either in one direction (Fig. 4.18(*a*)) or both (Fig. 4.18(*b*)), but it will be noticed that the intermediate lines are now lines of glide reflexion. These, as before, are shown in the diagrams by broken lines, continuous lines being used for simple reflexion. In the symbol *c11m*, the first '1' indicates absence of rotation and the second '1' absence of reflexion in lines perpendicular to the *x*-axis. For the second of these types the symbol can be abbreviated to *cmm*, since the reflexions in lines at right angles necessarily imply diad rotation. The numerous diad centres are shown in Fig. 4.19.

With glide reflexions occurring naturally in these two types it is hardly surprising that nothing new is obtained by splitting the element of pattern. Fig. 4.20 shows that, when this is done, the effect is to replace the glide lines by ordinary reflexion lines and vice versa. The pattern is still *c11m*.

We next consider the 'square' net (Fig. 4.21), noting first that there is no question of its being centred, as that would merely produce another square net with smaller unit cells, differently orientated (Fig. 4.22). With a square net there is a wider range of possible symmetries. The net has tetrad rotation about the vertices and centres of the squares and diad rotation about the mid-points of the sides (Fig. 4.23(*a*)). There are reflexion lines not only along the sides of the squares, with intermediate lines through the centres, but also along the diagonals, with intermediate glide lines (Fig. 4.23(*b*)). The complete symmetry

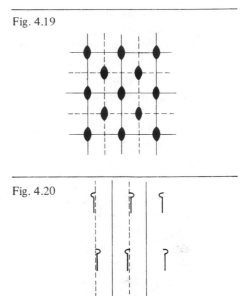

Fig. 4.19

Fig. 4.20

Fig. 4.21

Fig. 4.22

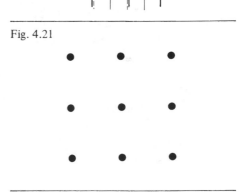

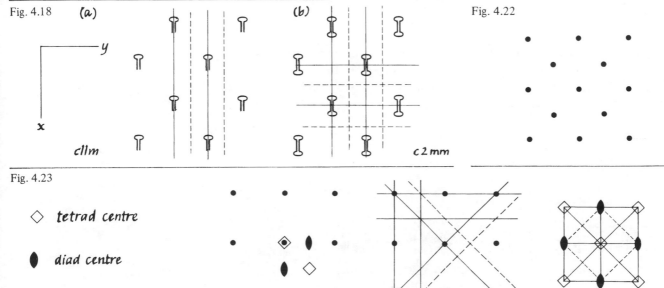

Fig. 4.18 (*a*)

y

x

c11m

(*b*)

c2mm

Fig. 4.23

◇ tetrad centre

◗ diad centre

(*a*)   (*b*)   (*c*)

chart for one unit cell (slightly enlarged) is shown in Fig. 4.23(*c*). Patterns of this type are shown in Fig. 4.24.

With tetrad rotation there cannot be any distinction between the axes of *x* and *y*. For this reason the first *m* in the symbol indicates reflexion lines in both these directions, and the second *m* refers to another pair of directions, those at 45° to the axes. With tetrad rotation we cannot have one set of reflexion lines without either reflexions or glide reflexions for the other set. (The theorem of p. 11 shows that, if the tetrad centres are at the intersections of the first set, there is a second set of reflexion lines at 45°. If they are not, the second set are glide lines.) We can, however, have tetrad rotation without any reflexions at all (Fig. 4.25(*a*)).

If we have glide lines through the centres of rotation (or reflexion lines not through those centres) we arrive at a pattern of the type shown in Fig. 4.25(*b*), with the pattern element reproduced at the centres of the squares. So we have a net of smaller squares, turned through 45°. There are ordinary reflexion lines, with intermediate glide lines, parallel to the sides of the original squares, but only glide lines parallel to those of the new squares. Turning the pattern through 45°, to bring the sides of the new squares parallel to the *x*- and *y*-axes, we have Fig. 4.26(*a*). The symbol for this type is 4*gm*, because in the directions parallel to the axes there are only glide lines, while in those at 45° there are ordinary reflexions. Further examples are shown in Figs. 4.26(*b*) and (*c*). The symmetry chart for this type is shown in Fig. 4.27(*b*), in contrast to that for *p*4*mm* (Fig. 4.27(*a*)).

There is one more kind of special parallelogram: the 60° rhombus, which divides into two equilateral triangles. The net based on this (Fig. 4.28) has hexad rotation about the vertices of the triangles, triad about their centres, and diad about the mid-points of the sides, as well as reflexion in the sides and altitudes of the triangles. We call this net *hexagonal*.

The motif superimposed on each point of the net may allow triad rotation (*p*3, Fig. 4.29) or hexad rotation (*p*6) without any reflexions; or with triad rotation it may have reflexions in the altitudes of the triangles (*p*3*m*1) or the sides (*p*31*m*); or with hexad rotation it may have reflexions in both these sets of lines.

To explain the notation we note first that the axes are oblique and placed as in Fig. 4.28. It is convenient to have a third axis as in Fig. 4.30. In the symbol, *m* in the second position after the prefix means that there is reflexion in lines perpendicular to the axes, i.e. in the altitudes of any of the triangles (Fig.

Fig. 4.24

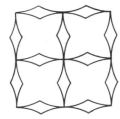

Fig. 4.25

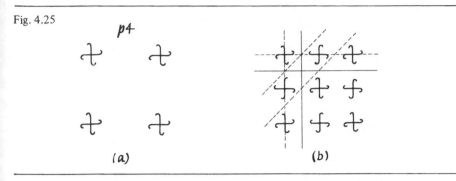

p4

(a)   (b)

Fig. 4.26

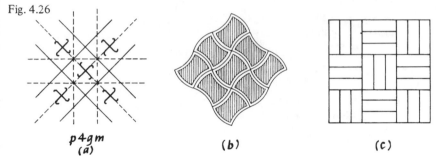

p4gm
(a)   (b)   (c)

Fig. 4.27

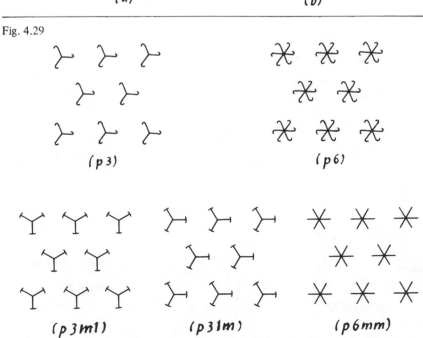

(*p4mm*)   (*p4gm*)
(a)   (b)

Fig. 4.29

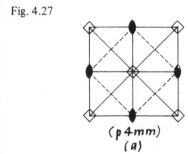

(*p3*)   (*p6*)

(*p3m1*)   (*p31m*)   (*p6mm*)

Fig. 4.28

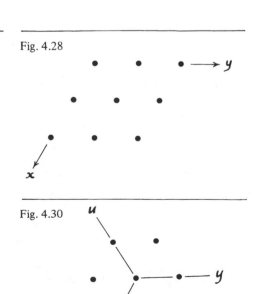

Fig. 4.30

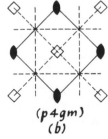

4.31(*a*)). These lines form angles of 60°, and the bisectors of those angles are the sides of the triangles (Fig. 4.31(*b*)). So *m* in the third position means reflexion in the sides of the triangles.* With hexad rotation, one *m* implies the other, but not so with triad. That is why there are two triad types, 3*m*1 and 31*m*, and only 6*mm* for hexad. The symmetries of the two triad types are contrasted in Fig. 4.32, one cell (enlarged) being shown in each case.

This concludes the list of the 17 types of wallpaper pattern. It will be noticed that the only rotations occurring in any of them are through angles of 360°/*n*, where *n* = 2, 3, 4 or 6. It is, for example, impossible to have a net allowing rotation through 360°/5, i.e. 72°. To prove this we note that in any net there is a minimum distance between points. Suppose then that, in Fig. 4.33, *A* and *B* are points of the net as near together as possible. The rotation property of the whole pattern demands that *C* and *D* should also be points of the net, and they are nearer together than *A* and *B*, contrary to our hypothesis. The same applies if *n* is greater than 6.

We give, in Fig. 4.34, some examples of different types of pattern. Further examples, for practice in identification, will be found at the end of Chapter 12.

Summarizing, there are 5 types of net and 17 types of wallpaper pattern. The 5 nets are shown in Fig. 4.35. The 17 wallpaper patterns are shown in Tables

*For the general rule, see p. 24.

Fig. 4.31

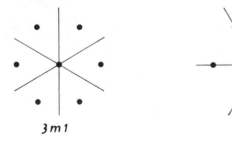

3*m*1   31*m*

Fig. 4.32

 ▲ *triad centre*

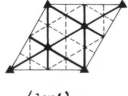 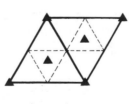

(3*m*1)   (31*m*)

Fig. 4.33

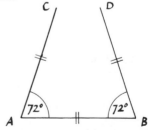

Fig. 4.35

Parallelogram   *Rectangular*   *Centred Rectangular*   *Square*   *Hexagonal*

Fig. 4.34

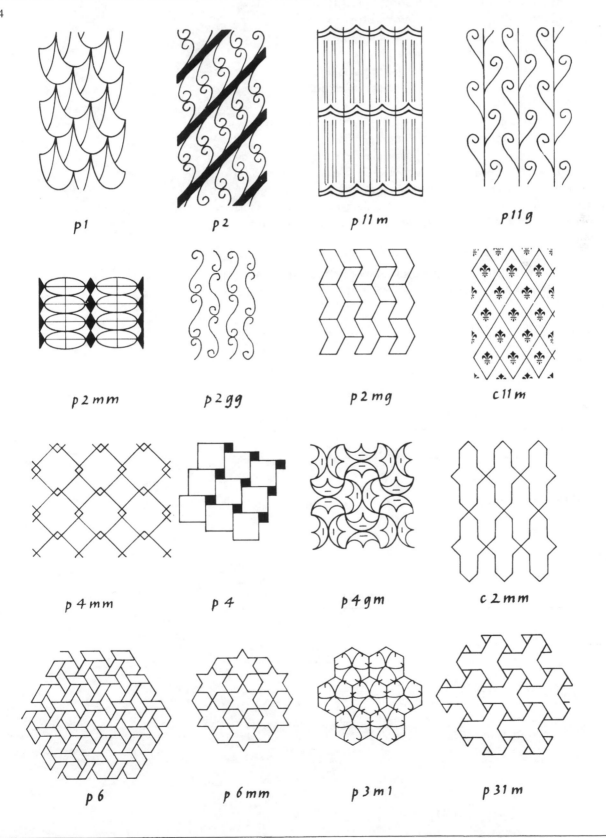

p1   p2   p11m   p11g

p2mm   p2gg   p2mg   c11m

p4mm   p4   p4gm   c2mm

p6   p6mm   p3m1   p31m

4.1 and 4.2 and diagrammatically in Fig. 4.36. Symbols are used as follows:

The prefix p or c denotes a primitive or centred net.

The numeral, in the first position after the prefix, shows the value of n for rotation through angles of $2\pi/n$.

In the second position there is m for mirror lines or g for glide lines (or 1 for neither) in a set of directions related by the rotation, one of them being perpendicular to the x-axis.

In the third position the same symbols are used for another set of related directions, intermediate to the former set.

Shortened forms, giving only the independent symmetries, are shown in parentheses. (These are commonly used by crystallographers, but they are of interest also to the designer as a guide to the amount of choice available. Thus with hexad rotation he may choose to have a set of reflexion lines at 60° intervals, but a second set follows automatically from the first.)

In Table 4.1 the reflexion or glide lines in each direction are a set of parallel lines spaced at repetition distances, as determined by the translations. There are also intermediate reflexion or glide lines, as shown in the lower line for each entry in Table 4.2.

Table 4.1

Net	Pattern types					
Paralellogram	$p1$	$p2$				
Rectangular	$p1m$ (pm)	$p1g$ (pg)	$p2mm$ (pmm)	$p2mg$ (pmg) or	$p2gm$ (pgm)	$p2gg$ (pgg)
Centred rectangular	$c1m$ (cm)	$c2mm$ (cmm)				
Square	$p4$	$p4mm$ $(p4m)$	$p4gm$ $(p4g)$			
Hexagonal	$p3$	$p3m1$	$p31m$	$p6$		$p6mm$ $(p6m)$

Table 4.2

Net	Pattern types					
Parallelogram	$p1$	$p2$				
Rectangular	$p1^{m}_{m}$	$p1^{g}_{g}$	$p2^{mm}_{mm}$	$p2^{mg}_{mg}$ or	$p2^{gm}_{gm}$	$p2^{gg}_{gg}$
Centred rectangular	$c1^{m}_{g}$	$c2^{mm}_{gg}$				
Square	$p4$	$p4^{mm}_{mg}$	$p4^{gm}_{gg}$			
Hexagonal	$p3$	$p3^{m}_{g}1$	$p31^{m}_{g}$	$p6$		$p6^{mm}_{gg}$

Fig. 4.36

Parallelogram net

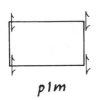
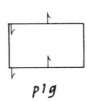

Rectangular net

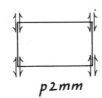
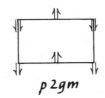
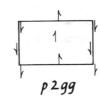

Centred rectangular net

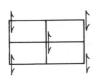

Square net

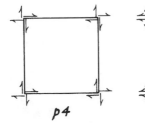
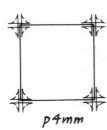
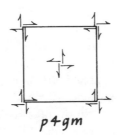

Hexagonal net

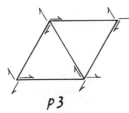
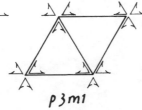
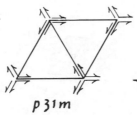
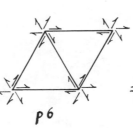
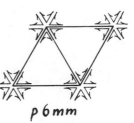

5

Finite objects in three dimensions

Many articles in common use exhibit symmetries of varying degree. A teapot, a chair or a spade show mirror symmetry, but as they are three-dimensional objects the reflexion is in a mirror plane rather than a mirror line. A table or a handbag may have reflexional symmetry in two mirror planes and will then also be self-coincident if given a half-turn about a vertical axis. Other objects, such as flower vases or chandeliers, may have more complicated symmetries, allowing reflexion in numerous planes and rotation through a variety of angles; and a sugar cube or a child's particoloured football allow rotation about many different axes.

In the movements suggested by these examples, whether reflexions or rotations, there is always at least one point of the object that remains fixed throughout, a centre point, or maybe the points of a central axis or mirror plane. The symmetry is called *point symmetry* and it will be proved in Appendix 1 that the symmetry of a finite object is necessarily of this kind. The movements cannot include translations, because a translation repeated indefinitely moves the object an infinite distance. In fact the only possible movements are reflexion in a plane through the fixed point, rotation about an axis through that point, and combinations of the same.

One such combination is inversion. In two dimensions this was the same as a half-turn, but in three dimensions it is not so, since inversion is then an opposite isometry, as illustrated in Fig. 1.06, p. 10. As mentioned on the same page, inversion in a point can be effected by a half-turn about any axis through the point followed by reflexion in the plane drawn through the point at right angles to that axis (or by the same two movements in the reverse order). This is illustrated in Fig. 5.01(*a*).

But inversion is only a special case of combined rotation and reflexion. It will be seen from Fig. 5.01(*b*) that a rotation through an angle a (from P_1 to P_2) followed by reflexion in a plane perpendicular to the axis of rotation (from P_2 to P_3) is the same as a rotation of $180°-a$ in the opposite direction (P_1 to

Fig. 5.01

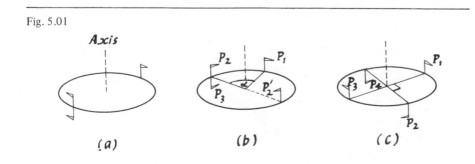

(a)   (b)   (c)

P_2') followed by inversion (P_2' to P_3). The combinations rotation–reflexion
and rotation–inversion thus amount to the same thing and it is not necessary to consider both. The modern practice is to use the latter, which we shall
refer to as *rotatory inversion*. The symbol $\bar{n}$ is used to indicate rotation through
$360°/n$ followed (or preceded) by inversion in a point on the axis of rotation.
Fig. 5.01(c) illustrates the movement $\bar{4}$ and its repetitions. With this convention, $\bar{2}$ is simple reflexion and $\bar{1}$ is inversion. (See Fig. 5.10).

We can now enumerate the types of point symmetry, considering only
rotations, reflexions and rotatory inversions. In two dimensions we found there
were two infinite sets of such types, built up from rotations with or without
reflexions. These same types occur in three dimensions, rotations about the
fixed point becoming rotations about an axis through that point, and reflexions
in lines through the point becoming reflexions in planes through the axis. Some
of these types are shown diagrammatically in Fig. 5.02.* The second set can be
illustrated by pyramids, as in Fig. 5.03. (Except for the first one they are all
right pyramids.) For the first set the same pyramids can be used if identical
unsymmetrical patterns are added to the faces, as in Fig. 5.04.

Two more infinite sets are obtained by adding reflexion in a plane perpen-

Fig. 5.02

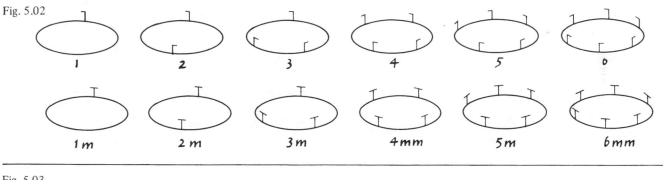

Fig. 5.03

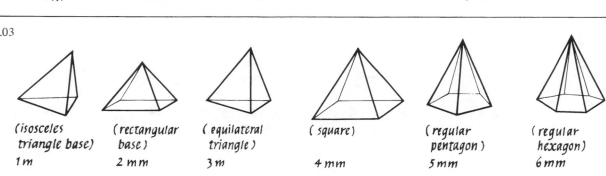

(isosceles triangle base)	(rectangular base)	(equilateral triangle)	(square)	(regular pentagon)	(regular hexagon)
1m	2mm	3m	4mm	5mm	6mm

*An alternative form of diagram, which some readers may prefer, is that used in
Symmetry Groups, by A.W. Bell & T.J. Fletcher.

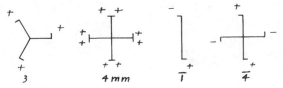

This system has the advantage of bearing a close resemblance to the stereograms
used in *International Tables* and later·in this book.

Fig. 5.04

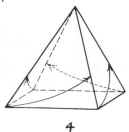

dicular to the axis of rotation (Fig. 5.05). The second of these sets, again, can be well represented by solids, this time prisms, on bases as before (Fig. 5.06). (Alternatively, double pyramids could be used.) For the first set (*m*, 2/*m*, etc.) a cyclic pattern needs to be added to the top face of each prism, with its reflexion on the bottom face (Fig. 5.07).

To explain the notation: there are three positions in the symbol, as shown in the second set above (Fig. 5.05). (In the first set the first position only is needed.) Of these three positions the first is used to indicate rotations about an axis (shown here as vertical and henceforward to be taken as the *z*-axis) and reflexion in a plane (the *xy*-plane) perpendicular to that axis. **It is to be noted that, throughout this book, we take the axes in the order *z*, *x*, *y*.** As before, numerals are used for rotations and the letter *m* for reflexions. If there are both rotation and reflexion the *m* is placed underneath. The second position, if there is one, refers to an axis or set of axes in the horizontal or *xy*-plane, these axes being related to one another by the rotation about the *z*-axis. An *m* in the second position indicates mirror planes perpendicular to those axes. The third position, if any, refers to another set of axes in the *xy*-plane, bisecting the angles formed by the first set, and to mirror planes perpendicular to them. Abbreviated forms are shown in parentheses.

Fig. 5.07

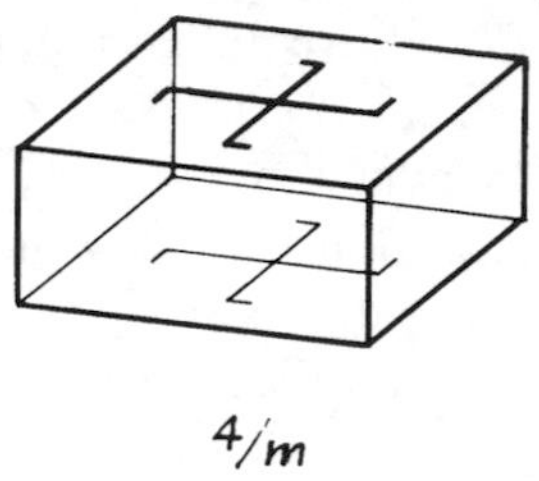

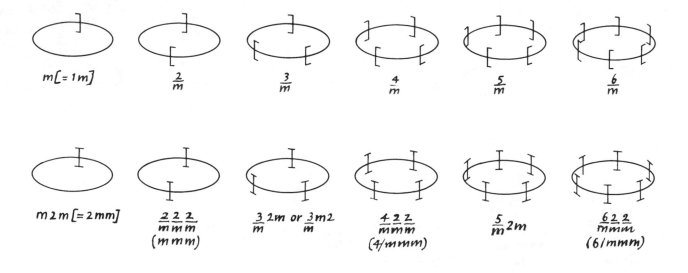

4/*m*

Fig. 5.05

$m[=1m]$ $\dfrac{2}{m}$ $\dfrac{3}{m}$ $\dfrac{4}{m}$ $\dfrac{5}{m}$ $\dfrac{6}{m}$

$m\,2\,m\,[=2mm]$ $\dfrac{2\ 2\ 2}{m\,m\,m}$ $\dfrac{3}{m}2m$ or $\dfrac{3}{m}m2$ $\dfrac{4\ 2\ 2}{m\,m\,m}$ $\dfrac{5}{m}2m$ $\dfrac{6\ 2\ 2}{m\,m\,m}$
                 $(m\,m\,m)$ $(4/mmm)$ $(6/mmm)$

Fig. 5.06

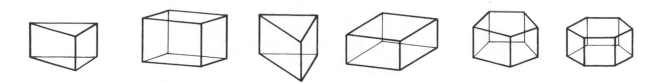

It may be noticed that there are two cases of overlap between members of these four sets of symmetry types. The patterns of type 1*m* in the second set and *m* in the third (Fig. 5.05) each show one reflexion and no other movement. They are therefore the same, though differently orientated. The *m* type is called a *second setting* of 1*m*. Again, 2*mm* in the second set is the same as *m*2*m* in the fourth, each containing two mirror planes, at right angles, and hence diad rotation about the line of intersection. We count *m*2*m* as a 'second setting' of 2*mm*.

Whereas intersecting mirror planes (as, for example, in the fourth set of symmetry types) necessarily imply rotations, it is possible to have rotation axes without any reflexions. There is thus a fifth set of point symmetry types with diad rotation axes in the *xy*-plane, as shown in Fig. 5.08.

There is again an overlap, the first one (12) in this set being the same as the second one (2) in the first set (Fig. 5.02). These types can be illustrated by adding diad rotation patterns to the side faces of the prisms in Fig. 5.06. An example is shown in Fig. 5.09.

The composite movement rotatory inversion has already been mentioned. The types built up from this are shown in Fig. 5.10. It will be seen that $\bar{1}$ represents pure inversion and $\bar{2}$ simple reflexion, the same as *m*. The symbol $\bar{6}$,

Fig. 5.08

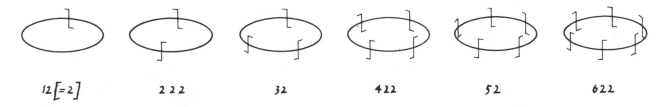

$12[=2]$ 222 32 422 52 622

Fig. 5.09

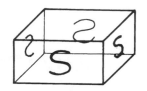

Fig. 5.10

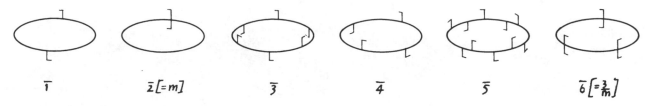

$\bar{1}$ $\bar{2}[=m]$ $\bar{3}$ $\bar{4}$ $\bar{5}$ $\bar{6}\left[=\frac{3}{m}\right]$

Fig. 5.13

$\overline{5}$

again, has the same meaning as $3/m$ and, if the series were continued, $\overline{10}$ would be the same as $5/m$. For reasons that will appear later, the symbol $\overline{6}$ is preferred to $3/m$.

Finally, we consider the effect of adding reflexions to the last set. To reflect in the horizontal (xy-) plane would not give anything new, but some new types appear if we reflect in planes through the vertical (z-) axis (Fig. 5.11). The first of these, $\overline{1}\frac{2}{m}$, is a new setting of $2/m$, and $\overline{2}m2$ is identical with $m2m$. The next three are new types, but $\overline{6}2m$ is identical with $\frac{3}{m}2m$ in the fourth set. As before, the symbol $\overline{6}2m$ (or $\overline{6}m2$) is preferred. The two diagrams in the lower row do not represent further types. They differ from those above them only in having the mirror planes differently orientated.

This last set of types, except for the first one, and those of the series $\overline{2}m2$, $\overline{6}m2$, $\overline{10}m2$, etc. can be well illustrated by the anti-prisms (Fig. 5.12), and the set shown in Fig. 5.10 by the same anti-prisms with diad rotation patterns on the side faces (Fig. 5.13).

There are thus seven infinite sets of point symmetry types and they correspond exactly to the seven types of frieze pattern (Fig. 5.14). They can be thought of as formed by wrapping the friezes round a cylinder, the particular member of each set depending on the number of repetitions in a circuit.

Fig. 5.14

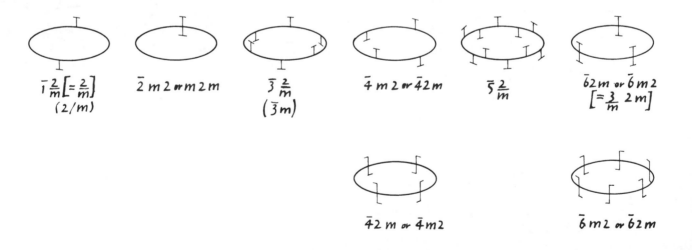

Fig. 5.11

$\overline{1}\frac{2}{m}\left[=\frac{2}{m}\right]$ $\overline{2}\,m\,2\,\text{or}\,m\,2\,m$ $\overline{3}\frac{2}{m}$ $\overline{4}\,m\,2\,\text{or}\,\overline{4}\,2\,m$ $\overline{5}\frac{2}{m}$ $\overline{6}2m\,\text{or}\,\overline{6}m2$
$(2/m)$ $(\overline{3}m)$ $\left[=\frac{3}{m}\,2\,m\right]$

$\overline{4}\,2\,m\,\text{or}\,\overline{4}\,m\,2$ $\overline{6}\,m\,2\,\text{or}\,\overline{6}\,2\,m$

Fig. 5.12

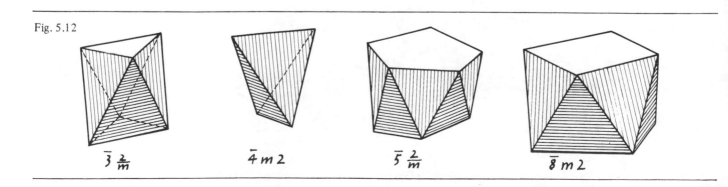

$\overline{3}\frac{2}{m}$ $\overline{4}\,m\,2$ $\overline{5}\frac{2}{m}$ $\overline{8}\,m\,2$

For use in repetitive patterns and for crystallographic purposes the only possible rotations are 1, 2, 3, 4 and 6. This would appear to give 35 types, but four of them are repetitions and four are re-orientations ('second settings'), leaving 27 genuinely different types.

It remains to consider the possibility of rotation axes other than the z-axis and those in the xy-plane. In a symmetrical pattern a rotation applies to every feature of the pattern, including other rotation axes. Suppose now we represent each rotation axis by the two 'poles' in which it cuts a sphere whose centre is the fixed point. If there is n-fold rotation about the z-axis ($n > 2$) and in addition m-fold rotation about some other axis, then the n-pole (on the z-axis) will be surrounded by a ring of m-poles. Moreover each m-pole will be surrounded by a ring of other m-poles. (If $m = 2$, this latter ring reduces to two poles on opposite sides of the original one, and thus a chain of 2-poles is formed along a great circle: in fact we have one of the cases already dealt with.) Except when $m = 2$, the m-poles are spread evenly over the surface of the sphere, in short at the vertices of a regular polyhedron. This means that the number of possibilities is limited by the fact that there are only five convex regular solids. (The Kepler–Poinsot polyhedra have their vertices arranged in the same way as the icosahedron or dodecahedron.) We have to consider (i) the tetrahedron, (ii) the cube and octahedron (these dual solids have the same symmetry), (iii) the icosahedron and dodecahedron (another pair of duals).

The tetrahedron can conveniently be considered as inscribed in a cube, as shown in Fig. 5.15. The axis shown here as vertical (the z-axis) is seen to be an axis of tetrad rotatory inversion ($\bar{4}$), and there are three axes of this kind, as well as four triad rotation axes along the diagonals of the cube. There are also six reflexion planes (the diagonal planes of the cube). The symbol for this type of symmetry is thus $\bar{4}3m$. (Note that the $\bar{4}$ axes are related to each other by the triad symmetry and the triad axes by the $\bar{4}$ symmetry.)

The pattern need not have the full symmetry of the tetrahedron. If the indirect movements (opposite isometries) are excluded the $\bar{4}$ axis becomes a diad and we then have three diad axes and four triad. The symbol for this type is 23 and it may be illustrated by placing a C_3 (rotation 3) pattern on each face of the tetrahedron (Fig. 5.16).

The cube and octahedron have three tetrad rotation axes, four triad-inversion and six diad. One of each is shown in Fig. 5.17. The triad-inversion includes simple inversion and this combines with any half-turn to give a reflexion. The consequence is that there are nine reflexion planes, perpendicular respectively to the six diad axes and the three tetrads. The full symbol for this type of symmetry is $\frac{4}{m}\bar{3}\frac{2}{m}$, but it is commonly abbreviated to $m3m$.

Polyhedra such as those of Fig. 5.18, obtained by cutting off the eight

Fig. 5.15

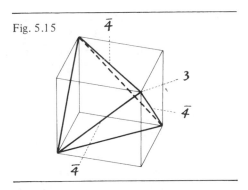

Fig. 5.16

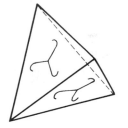

Fig. 5.17

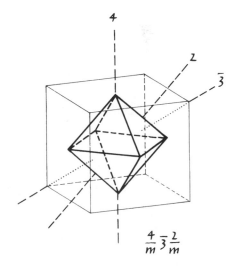

Fig. 5.18

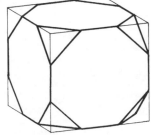 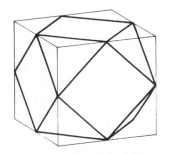 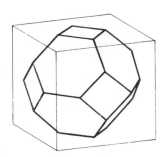

corners of a cube, have this same symmetry. They are in fact intermediate between the cube and octahedron. If the twelve edges of the cube are sliced off, as in Fig. 5.19, the resulting solids again have the same symmetry as the cube. (Fig. 5.19(b) shows the rhombic dodecahedron.) But if only four corners are cut off, as for the tetrahedron, the symmetry is reduced.

This cubic symmetry can be reduced in two ways. The first is by excluding the indirect movements, leaving the three tetrad rotations, four triads and six diads. The symbol is thus 432. This type can be illustrated by placing identical C_4 patterns on all the faces of the cube (Fig. 5.20).

The other way is to reduce the tetrad symmetry to diad, keeping the triad rotations and the reflexions in planes parallel to the faces of the cube. This is shown in Fig. 5.21. It will be noticed that the triad rotatory inversion of Fig. 5.17 (including simple inversion) is retained, but not the diad rotation. The symbol is $\frac{2}{m}\bar{3}$, commonly abbreviated to $m3$. This kind of symmetry is found in crystals of iron pyrites.

The icosahedron and dodecahedron each have six axes of pentad rotation, ten triad, and fifteen diad. Figs. 5.22(a) and (b) show these solids viewed along the pentad and triad axes respectively. They also have inversion, with the consequences that the pentad and triad axes are axes of rotatory inversion and that

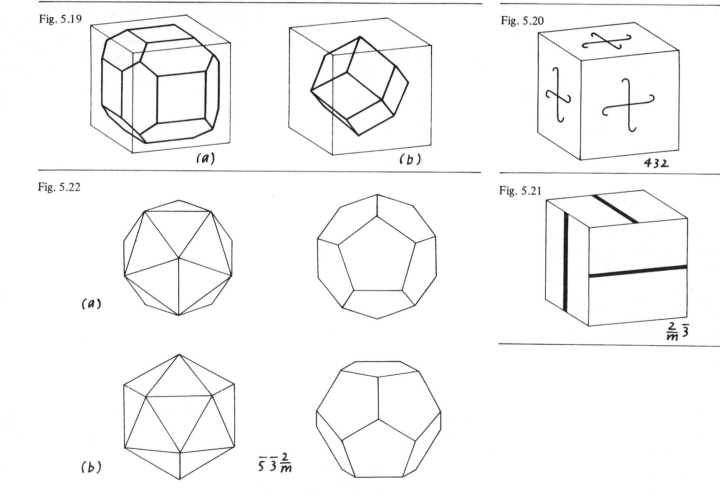

Fig. 5.19

(a)   (b)

Fig. 5.20

432

Fig. 5.22

(a)

(b) $\bar{5}\,\bar{3}\,\frac{2}{m}$

Fig. 5.21

$\frac{2}{m}\bar{3}$

Fig. 5.23

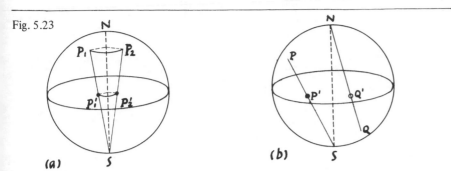

(a)   (b)

there are 15 reflexion planes perpendicular to the diad axes. The symbol for this type is $\bar{5}3\frac{2}{m}$.

If identical cyclic patterns are placed on each of the faces of either of these solids the inversion symmetry is destroyed but the rotations are still possible. The symbol is then 532.

These last two types are not of interest to crystallographers because pentad rotation is not possible for a pattern that repeats in more than one direction. The total number of crystallographic types of point symmetry is thus 27 + 5, i.e. 32.

To summarize this chapter and to present the 32 crystallographic point symmetry types in a form convenient for reference, we resort to stereograms, as used in some other books, notably in *International Tables for X-ray Crystallography*. It is supposed that the symmetry movements are represented on the surface of a sphere, by an arbitrary point P_1 and further points $P_2, P_3 \ldots$ in all the positions it might move to (Fig. 5.23(a)). Points P_1, P_2 in what may conveniently be described as the 'northern' hemisphere are projected from the 'south pole' into corresponding points P_1', P_2' in the equatorial plane and are represented by dots in a circular diagram representing that plane. Similarly points Q in the 'southern' hemisphere are projected from the 'north pole' into points Q' in the same plane and represented by rings in the same diagram (Fig. 5.23(b)). Positions reached by 'opposite' movements (e.g. reflexions or inversion) are shown in red.

Thus, for example, tetrad rotation about the north–south axis (the z-axis) will be represented as in Fig. 5.24(a), and diad rotation about an axis AB in the equatorial plane will be shown as in (b). Reflexions in the equatorial plane and in a plane containing the z-axis are shown in (c) and (d) respectively, and inversion in (e).*

The 32 types, and the four 'second settings', are shown in this manner in Fig. 5.25. The full symbols are shown (axes in the order z, x, y), with shortened forms in parentheses underneath. For second settings reference is given in brackets alongside. For the four identical pairs the diagram is omitted for one of each pair. The alternative orientations, $\bar{4}2m$ for $\bar{4}m2$ and $\bar{6}2m$ for $\bar{6}m2$ are shown in Fig. 20.10, p. 130.

> *It is important that the reader should understand this form of diagram. An alternative, perhaps easier, interpretation, which serves well enough for all except the last five diagrams of Fig. 5.25, is to think of the circles as discs, with the dots representing points on one side, and the rings points on the other. Expressing it algebraically, if the black dots in the diagrams (b), (c), (d) and (e) represent the point (x, y, z), the black ring in (b) is $(x, -y, -z)$, the red ring in (c) is $(x, y, -z)$, and the red dot in (d) is $(x, -y, z)$. The red ring in (e) is $(-x, -y, -z)$.

Fig. 5.24

y

x

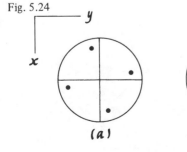
(a)

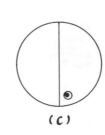
A
B
(b)

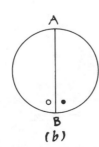
(c)

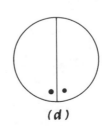
(d)

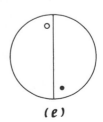
(e)

Fig. 5.25

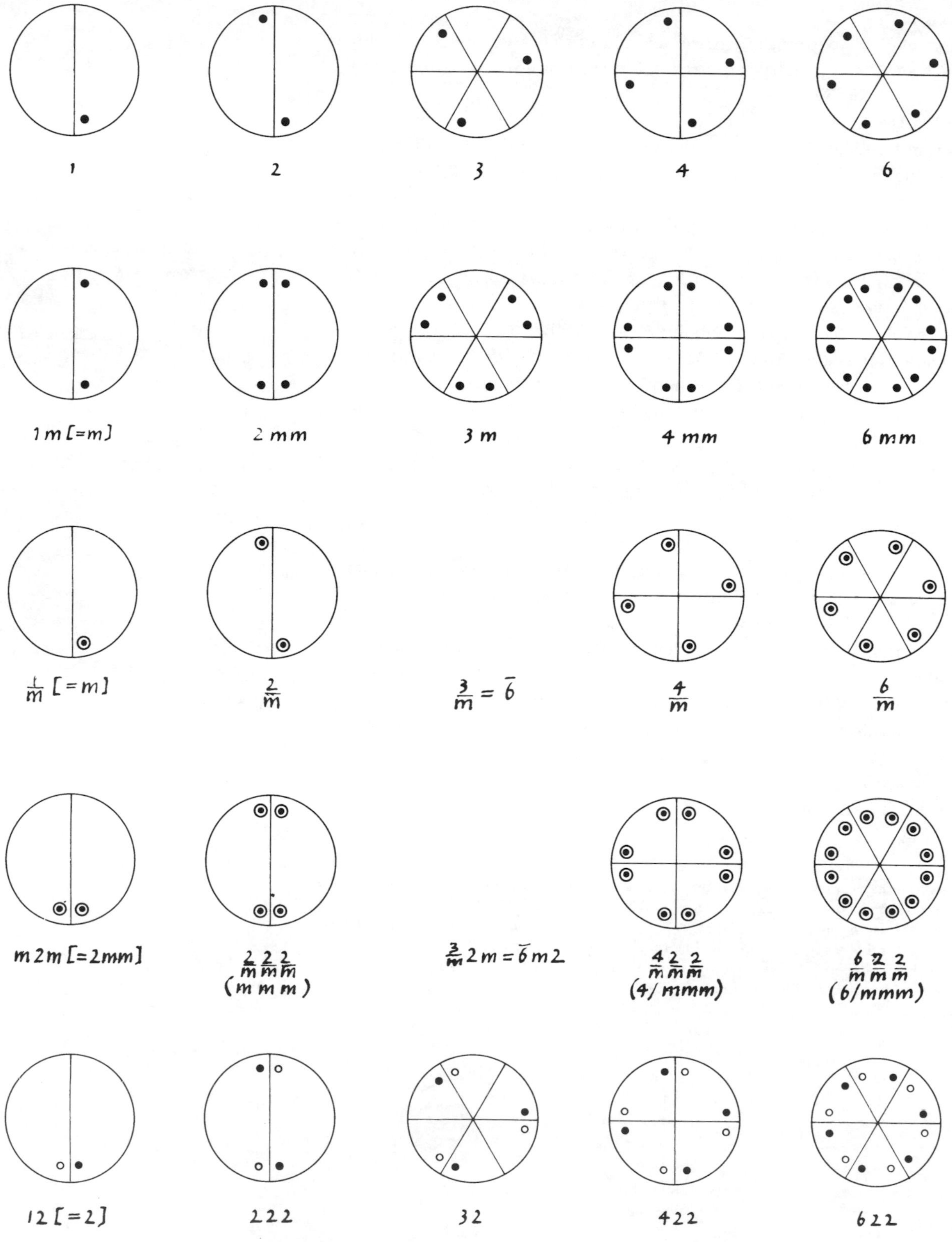

| 1 | 2 | 3 | 4 | 6 |

| $1m\,[=m]$ | $2\,mm$ | $3\,m$ | $4\,mm$ | $6\,mm$ |

| $\dfrac{1}{m}\,[=m]$ | $\dfrac{2}{m}$ | $\dfrac{3}{m}=\bar{6}$ | $\dfrac{4}{m}$ | $\dfrac{6}{m}$ |

| $m\,2m\,[=2mm]$ | $\dfrac{2}{m}\dfrac{2}{m}\dfrac{2}{m}$ (mmm) | $\dfrac{3}{m}2m=\bar{6}m2$ | $\dfrac{4}{m}\dfrac{2}{m}\dfrac{2}{m}$ $(4/mmm)$ | $\dfrac{6}{m}\dfrac{2}{m}\dfrac{2}{m}$ $(6/mmm)$ |

| $12\,[=2]$ | 222 | 32 | 422 | 622 |

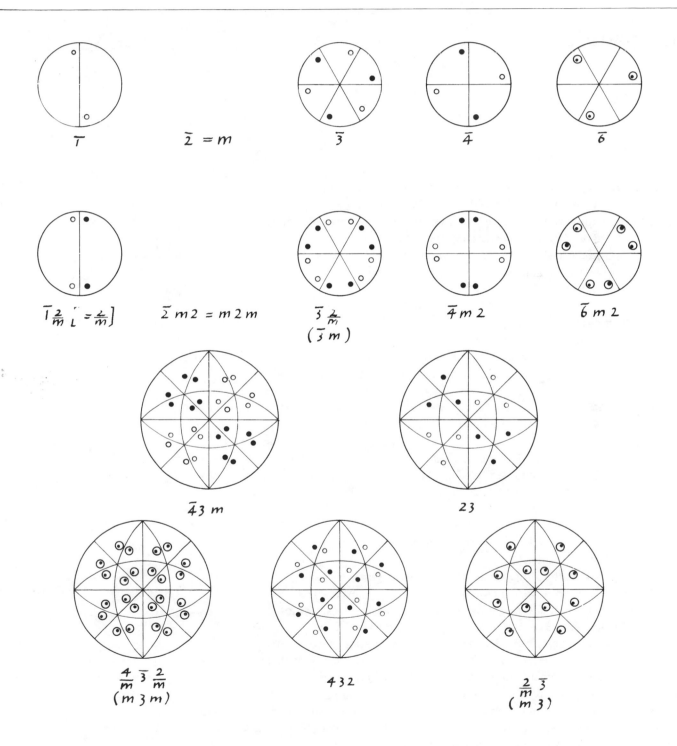

6

Rod patterns

A line pattern can be built up as well from a three-dimensional motif as from a two-dimensional one. We may proceed from a flat frieze to one carved in relief; thence to one with a front and a back (a 'two-sided ribbon' pattern); and finally to a 'rod pattern' formed by repeating any three-dimensional motif at regular intervals along a line. The symmetries of a rod pattern are of course limited by those of the row of points on which it is built. These are:

> inversion,
> rotation of any amount about the longitudinal axis,
> half-turns about transverse axes,
> reflexions in transverse planes,
> reflexions in planes containing the longitudinal axis.

Taking the line of the row as z-axis we may use a three-dimensional motif of any of the types in the seven infinite sets described in Chapter 5. Their symmetries, being also symmetries of the row, will be symmetries of the pattern as a whole. But if we use a motif based on one of the regular polygons some of the rotations will not apply to the whole pattern. Thus a row of clothes pegs could have the same symmetries as an individual peg, but a row of cubical blocks cannot have all the symmetries of a cube.

For a two-sided ribbon the choice is much more limited, because half-turns are the only admissible rotations. There are in fact only 31 types. (They correspond exactly to the 31 types of bicoloured frieze shown in Fig. 27.07. p. 199, with white and black representing front and back respectively and grey indicating both front and back.)

In forming a rod pattern there is no need to restrict ourselves to the crystallographic rotations, with n equal to 1, 2, 3, 4 or 6; nevertheless in enumerating the types it is very convenient to do so, to avoid a surfeit of infinities. We then have 31 types of motif available, as shown in Fig. 5.25, the five polyhedral types being excluded. (The 31 include the four second settings.) Simple repetition of these gives 31 types of line pattern. With many of them, however, the pattern element can be split to produce glides or screws, a *screw* being a combination of a rotation about an axis with a translation along that axis as, for example, in a 'spiral' staircase. In fact it is found that there are altogether 75 types. They will be enumerated in Part II: for the present we give some examples to show the variety of possibilities.

The post with projections (Fig. 6.01(a)), as sometimes seen on cliffs for use with rocket life-saving apparatus, has diad rotation symmetry about the longitudinal axis and two reflexion planes through it. Its pattern element has the symmetry $2mm$, and the symmetry of the post itself (supposed infinitely long) is $r2mm$.* If, however, a post were made with the projecting pieces arranged as

*The axes are taken, as before, in the order z, x, y, the z-axis being in the longitudinal direction.

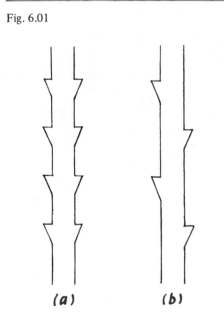

Fig. 6.01

(a)　　　(b)

steps (Fig. 6.01(*b*)), there would be a glide reflexion instead of the simple reflexion and a screw instead of the rotation. The symmetry would be $r2_1mg$, where 2_1 indicates a half-turn screw and g a glide.

To represent these symmetry types diagrammatically, in the manner of Fig. 5.25, one diagram is sufficient for the first kind of post, as it has simple repetition, but for the second kind we need two, to show successive cross-sections (Fig. 6.02). The diagrams are drawn with the longitudinal axis as z-axis.

In Fig. 6.02(*b*) one reflexion has become a glide and the diad rotation has become a screw, i.e. the half-turn is accompanied by a move forward along the axis of rotation. (The notation 2_1 for this is explained below.) But glides and screws do not always go together. Fig. 6.03(*a*) shows a wooden fence with supports on one side. This has transverse reflexion, but no other symmetry. If the posts were places alternately on one side and the other (Fig. 6.03(*b*)) it would have glide reflexion as well. The diagrams, again taking the longitudinal axis as z-axis, are as shown in Fig. 6.04. Another example of a glide reflexion without a screw is a double row of bicycles stacked in the way often seen outside factories.

For a screw without a glide, nature provides some good examples in the plants that send out shoots in regularly varying directions at regular intervals up the stalks. For a more accurate example consider the steps of a so-called 'spiral'

Fig. 6.02

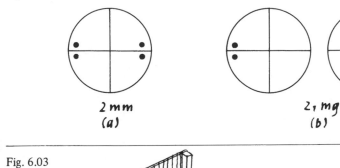

2 mm

(a)

2_1 mg

(b)

Fig. 6.03

(a)

(b)

Fig. 6.04

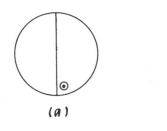

(a)

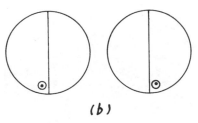

(b)

staircase, or the windows in a stone tower containing such a staircase. It is to be noted that the staircase can wind either right-handedly or left-handedly. Patterns so related are called *enantiomorphic*.

A paper chain, as made by children for Christmas decorations (Fig. 6.05), allows a quarter-turn screw, as well as half-turns about transverse axes. It also has transverse reflexion planes and, in the longitudinal direction, two reflexion planes with intermediate glide planes. The diagram is shown in Fig. 6.06, with the longitudinal reflexion and glide planes marked by continuous and broken lines respectively.

For screws a new notation is needed. Every screw can be regarded as formed from a rotation pattern by starting with one sub-cell of the pattern and moving it forward as it is rotated. Fig. 6.07 represents a screw in which there is a triad turn combined with an advance of $\frac{1}{3}$ of a unit (taking the repetition distance as 1 unit). The three circles in this diagram represent successive cross-sections, moving along the z-axis in the upward, or positive, direction. The screw is thus right-handed, and we count this as a turn about the z-axis in the positive sense. This screw is denoted by 3_1, the 3 indicating, as before, a triad turn, and the suffix 1, with the 3 used again as denominator, giving the fraction of a unit for the forward movement. A similar but left-handed screw (Fig. 6.08) could logically be denoted by 3_{-1}, but more conveniently by 3_2, meaning that a

Fig. 6.05   Fig. 6.06

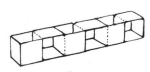

Fig. 6.07

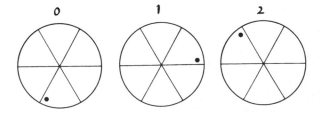

Fig. 6.08

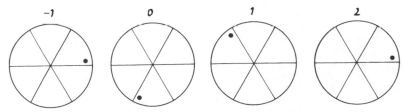

triad turn (in the positive sense) accompanies an advance of $\frac{2}{3}$ of a unit (in the positive direction). Fig. 6.08 shows that a forward movement of $\frac{2}{3}$ of a unit gives the same cross-section as a backward movement of $\frac{1}{3}$. (In mathematical terms, 2 is congruent to -1, mod 3.)

The screws represented by 4_1, 4_2 and 4_3 are shown in Fig. 6.09. In 4_2 the advance for a quarter-turn is $\frac{2}{4}$, i.e. $\frac{1}{2}$ of a unit. This makes a half-turn for 1 unit, and as there is also simple repetition at unit interval this type necessarily includes diad rotation.

Using this notation for screws, the symbols for the patterns mentioned so far are as follows:

the posts (Figs. 6.01, 6.02)	(a) $r2mm$,
	(b) $r2_1mg$;
the fences (Figs. 6.03, 6.04)	(a) rm,
	(b) $rm2g$;
the paper chain (Figs. 6.05, 6.06)	$r\frac{4_2}{m}\frac{2}{m}\frac{2}{g}$

It will be remembered that the first position in the symbol refers to rotation about the z-axis (longitudinal) and to reflexion in planes perpendicular to it; the second to a set of transverse axes related by the said rotation (this means only one axis when the rotation figure is 2, but two when it is 4), and to planes perpendicular to them; and the third to another set of axes and planes, bisecting the angles formed by the previous set.

Fig. 6.10 shows a pattern to be seen on a pillar in the church at Kirkby Lonsdale in Cumbria. This shows a screw rotation 12_6 and there is a set of six reflexion planes related by the 12-fold rotation, with intermediate glide planes. The symbol is $r\frac{12_6}{m}\frac{2}{m}\frac{2}{g}$.

At the chateau of Chambord there is a double staircase (Fig. 6.11) showing a left-handed screw in an octagonal setting. The symbol is $r8_6$. There are no reflexions.

Fig. 6.09

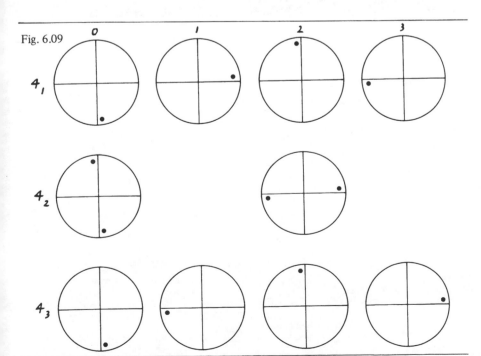

Fig. 6.10

Fig. 6.11

7

Layer patterns

In two dimensions a pattern based on repetition in two directions was called a 'wallpaper pattern'. If a three-dimensional motif is used in this way we have something more than a wallpaper and there are many more than 17 possibilities. We call it a '*layer pattern*', or 'a plane pattern in three dimensions'. Any one of the 31 types of pattern element mentioned in the last chapter can be used, this time with one of the five nets. Simple repetition in this manner provides 42 types of layer pattern; but glides and screws may also be used, giving many further types, a total of 80 in all. First we give some examples of simple repetition.

The seats in a concert hall, arranged with one seat behind another (Fig. 7.01(a)), show a motif of type $1m$ placed on a rectangular net; but if they are arranged with each seat behind a gap in the row in front (Fig. 7.01(b)) the net is centred. The notation for these patterns is $p11m$ and $c11m$, the prefixes p and c showing whether the net is primitive or centred, the two '1's indicating that there is no rotation about either the z-axis or the x-axis, nor reflexions in planes perpendicular to those axes. The m in the third position indicates reflexion in planes perpendicular to the y-axis. In the diagrams of Fig. 7.02 we show one cell of the net for each type, with dots for corresponding points of the pattern.

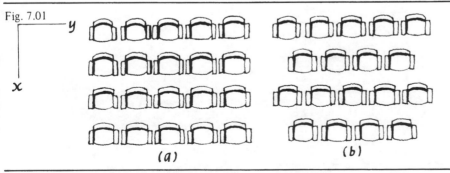

Fig. 7.01

y

x

(a)   (b)

Fig. 7.02

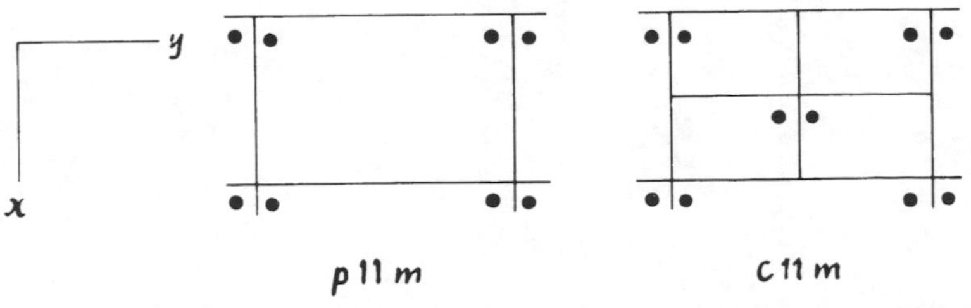

y

x

$p\,11\,m$ $c\,11\,m$

Next consider the pattern of the radiator shown in Fig. 7.03(*a*), the back having the same pattern as the front. The pattern element has the point symmetry $\frac{2}{m}\frac{2}{m}\frac{2}{m}$, and the whole pattern is described as $p\frac{2}{m}\frac{2}{m}\frac{2}{m}$. The diagram is shown in Fig. 7.03(*b*).

The trellis-work shown in Fig. 7.04(*a*) has no reflexions, but it allows diad rotation about axes in all three directions, so its symbol is *c*222. The diagram for this symmetry is shown in Fig. 7.04(*b*).

Eggs placed upright in a one-layer box with square partitions (Fig. 7.05(*a*)) form a pattern of type *p*4*mm*, but if laid on their sides in the same box (Fig. 7.05(*b*)) the symmetry is *pm*2*m*. Diagrams are shown in Fig. 7.06. The net in

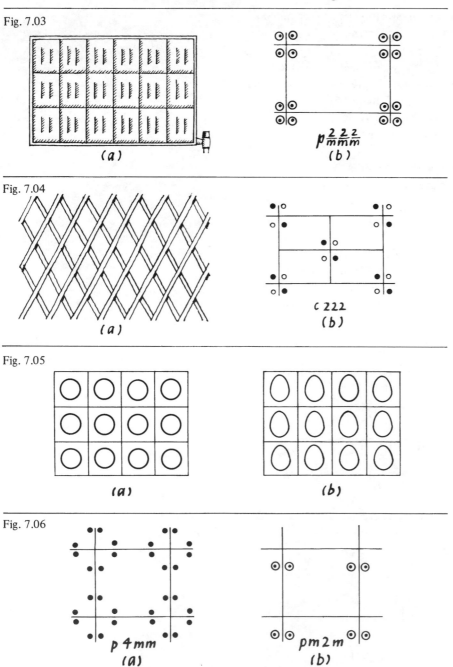

Fig. 7.03

$p\frac{2}{m}\frac{2}{m}\frac{2}{m}$

(a)   (b)

Fig. 7.04

c 222

(a)   (b)

Fig. 7.05

(a)   (b)

Fig. 7.06

p 4*mm*   *pm* 2*m*

(a)   (b)

Fig. 7.06(*b*) might just as well be rectangular as square. In fact both the square net and the pattern element have higher symmetries than the pattern as a whole. The symmetry of the whole pattern is, as it were, the highest common factor of the other two.

As with wallpaper patterns, a centred square reduces to a smaller primitive one turned through an angle of 45°. The embossed pattern shown in Fig. 7.07(*a*) is *p4mm*; and if the smaller circles are enlarged to equal the size of the others it is still *p4mm*, but with the square net smaller and turned through 45° (Fig. 7.07(*b*)).

We next consider glides and screws. The only possible screws are half-turn screws about axes in the *xy*-plane. For glides it is now necessary to specify the reflexion plane as well as the direction of the translation. In a layer pattern the translation must be in the *xy*-plane, but the reflecting plane may be either the *xy*-plane itself or one perpendicular to it. The second of these possibilities is illustrated in Fig. 7.07, where there are in each case glide planes intermediate to one set of reflexion planes. Examples of the first possibility are shown below.

The direction of a glide is indicated in the symbol by using *a* for a glide in the direction of the *x*-axis, *b* for one in the direction of the *y*-axis, *g* for both together, and *n* for one in the diagonal direction. The reflecting plane is shown by the position of such letters in the symbol, the first position after the prefix referring, as usual, to the *z*-axis and planes perpendicular to it. (The *xy*-plane is of course one of these and may conveniently be called the 'Z-plane'. Thus the first position refers to the *z*-axis and the Z-plane.)

Consider now the basket-work pattern shown in Fig. 7.08. It is based on a centred rectangular net. There is diad rotation about axes in the *z*-direction, and in the Z-plane there are glides in both the *x*- and *y*-directions (first symbol 2/*g*). There is also diad rotation about the *x*- and *y*-axes and reflexions in planes perpendicular to them. So the full symbol is $c\frac{2}{g}\frac{2}{m}\frac{2}{m}$ and the diagram is as shown in the figure.

Fig. 7.07

(a)   (b)

Fig. 7.08

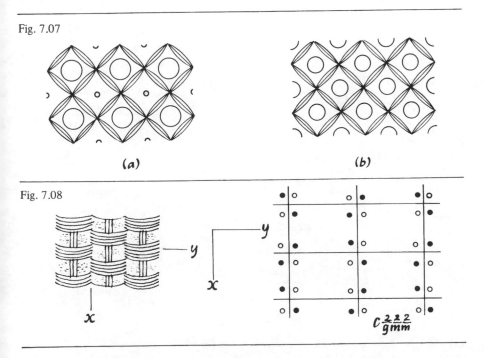

A sheet of corrugated iron allows infinitesimal translations in one direction (see Chapter 9) but if it is divided into sections, as in Fig. 7.09, we have a plane symmetry pattern whose symbol is $p\frac{2_1}{b}\frac{2}{m}\frac{2}{m}$.

Tetrad rotatory inversion, as illustrated by the two sticks in Fig. 0.02 (p. 1), appears again in the simple cross-stitch pattern of Fig. 7.10(a). (We ignore the canvas and the fact that the stitches turn in towards it.) The pattern is based on a square net, with diad rotation axes parallel to the sides of the squares and mirror planes in the diagonal directions. The symbol is $p\overline{4}2m$. If, however, the stitches are arranged as in Fig. 7.10(b), the net, though still square, is turned through 45° and there is tetrad rotation. In Fig. 7.10(c) we show a unit cell of this pattern (enlarged). The tetrad axes at the corners have identical environments and there is a fifth tetrad axis in the centre with the same environment reflected in the xy-plane. We therefore have $4/n$ in the first position in the symbol. As shown in the figure, there are screw axes parallel to the sides of the squares and rotation axes parallel to the diagonals, with mirror planes perpendicular to both pairs of directions. So the full symbol is $p\frac{4}{n}\frac{2_1}{m}\frac{2}{m}$.

In the wire mesh of Fig. 7.11(a) we again have $4/n$, a unit cell with its tetrad axes being as shown in Fig. 7.11(b). There are diad rotations in both pairs of directions, with glide reflexions parallel to the sides of the squares, as shown. The symbol is $p\frac{4}{n}\frac{2}{g}\frac{2}{m}$.

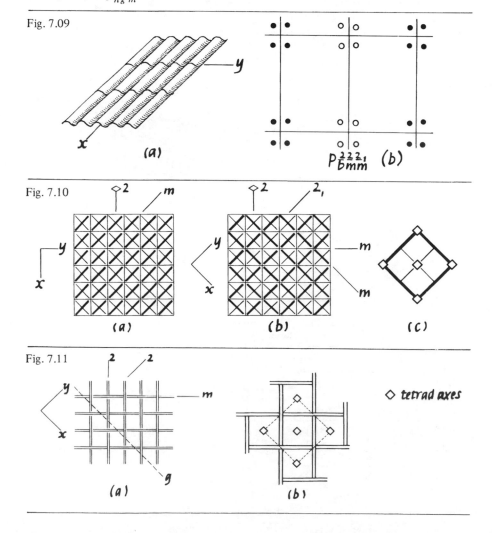

Fig. 7.09

$P\frac{2}{b}\frac{2}{m}\frac{2_1}{m}$ (b)

(a)

Fig. 7.10

2   m   2   2₁

(a)   (b)   (c)

m   m

Fig. 7.11

2   2

m

g

(a)   (b)

◇ tetrad axes

The honeycomb provides an interesting example, the hexagonal compartments on one side being placed opposite vertices of the hexagons on the other side (Fig. 7.12). There is triad inversion symmetry about axes in the z-direction (passing through vertices of hexagons on both sides) and diad rotation about axes as shown in the figure, with mirror planes perpendicular to them. It might at first seem that there ought to be glide reflexions in the direction of the diad axes and screws about axes parallel to the mirror planes, but on close examination it will be found that this is not so. The symbol for this pattern is $p\,\bar{3}\frac{2}{m}$ and the diagram is as shown in Fig. 7.13.

The 80 types of layer pattern will be enumerated in Part II (Chapter 22).

Fig. 7.12

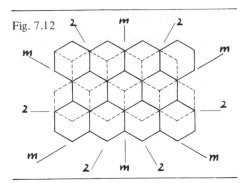

Fig. 7.13

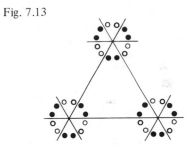

8

Space patterns

Eggs are commonly packed (or stacked) with each layer arranged according to
a square net, successive layers being so placed that each egg is above the centre
of a square formed by four eggs in the layer below (Fig. 8.01). Starting from a
single egg, the whole pattern can be built up by applying two horizontal translations T_1 and T_2, and a third translation T_3 inclined obliquely to the horizontal plane. With any point as origin, T_1 and T_2 determine a net, and T_3
repeats the net in parallel planes. This forms a *lattice*. It represents the mode of
repetition of the pattern element.

We may take any one point of the pattern as origin (e.g. one corner of one of
the trays in which the eggs are placed), with axes of x, y and z in the directions
of the translations and the sides of the unit cell as units in those directions. The
points of the lattice are then all those points with integral coordinates. Alternatively, in Fig. 8.01, we may prefer to take the z-axis perpendicular to the plane
of the net and to regard the lattice cell as body-centred.

The types of space pattern are limited by the types of lattice and so we
consider first the various possible lattices, as distinguished by their symmetries.
The unit cell is a parallelepiped and it may sometimes be centred on one pair of
faces, or on all faces, or at its body-centre. It cannot be centred on two pairs of
faces without the third. To see this, note that successive diagonal movements
on two faces are equivalent to one on the third face (Fig. 8.02(*a*)) and hence
successive half-diagonal movements on or parallel to the same two faces are
equivalent to a half-diagonal movement on the third (Fig. 8.02(*b*)). (In math-
ematical terms, the sum of the vectors $(-\frac{1}{2}, 0, \frac{1}{2})$ and $(\frac{1}{2}, \frac{1}{2}, 0)$ is $(0, \frac{1}{2}, \frac{1}{2})$,
representing a movement from the origin to the centre of a third face.)

The lattice whose unit cell is the general parallelepiped is called *triclinic*,
because all three angles between the axes are oblique (Fig. 8.03). The only
symmetry of the unit cell is inversion in the centre, but the lattice as a whole
has inversion in the lattice points and in the mid-points of the lines joining
them. (This includes the mid-points of the edges of the cells, the centres of the

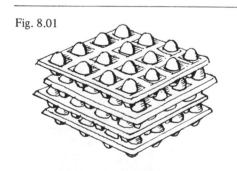

Fig. 8.01

Fig. 8.02

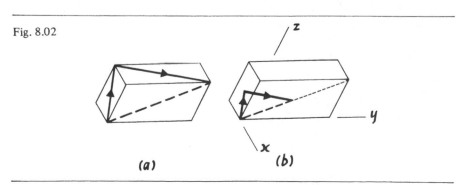

(a)   *(b)*

Fig. 8.03

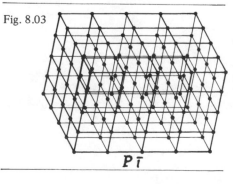

$P\bar{1}$

faces, and the body-centres.) The symbol is $P\bar{1}$, indicating a primitive lattice with inversion symmetry. (A centred lattice would reduce to a primitive one by suitable choice of axes.)

For the lattice to have other symmetries the unit cell must be one of the various special parallelepipeds, and these we now consider. Using the general parallelogram net as base layer, the third translation T_3 may be perpendicular to the plane of the net. (No extra symmetries are introduced by having it perpendicular to one but not both of T_1 and T_2.) The lattice now has half-turn rotation about the z-axis and numerous parallel axes. It also has reflexion in the xy-plane (and parallel planes). This lattice is called *monoclinic*, because one of the three angles between the axes is oblique. Its symmetry may be expressed as $2/m$ (diad rotation about the z-axis and reflexion in planes perpendicular to that axis). The unit cell may be primitive, or centred on one pair of rectangular faces. (There is no need to consider the possibility of its being body-centred or centred on all faces, because in such cases a different choice of parallelogram for the basic net would reduce the lattice to the type with unit cell centred on one pair of faces only. This is shown in Fig. 8.04, where the unlabelled dots are lattice points in the xy-plane and those with $\frac{1}{2}$ attached are displaced $\frac{1}{2}$ unit in the z-direction.) For notation we use prefixes C, A, B for lattices centred on faces perpendicular to the z-, x-, y-axes respectively, F for lattices centred on all faces, and I for body-centred. The two forms of monoclinic lattice are thus $P2/m$ and $A2/m$ (or $B2/m$). In Fig. 8.05 we show a unit cell of each.

With a rectangular net (in the xy-plane) as base, the third translation (and the z-axis) must be perpendicular to the plane of the net, if we are to avoid types already dealt with. This is the *orthorhombic* lattice, with rectangular axes of z, x, y. The symmetries are $\frac{2}{m}\frac{2}{m}\frac{2}{m}$, often abbreviated to *mmm*. This lattice may be centred on one pair of faces of the unit cell (C) or all faces (F) or it

Fig. 8.04

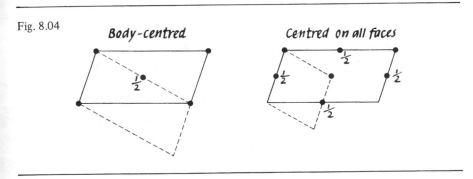

Fig. 8.05

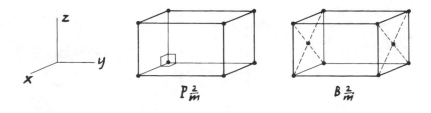

Fig. 8.06

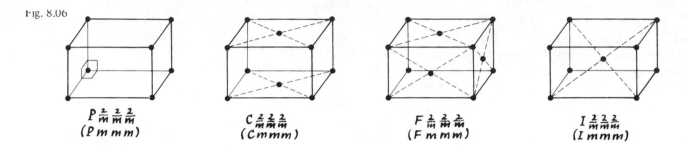

$$P\frac{2}{m}\frac{2}{m}\frac{2}{m}$$
$$(Pmmm)$$

$$C\frac{2}{m}\frac{2}{m}\frac{2}{m}$$
$$(Cmmm)$$

$$F\frac{2}{m}\frac{2}{m}\frac{2}{m}$$
$$(Fmmm)$$

$$I\frac{2}{m}\frac{2}{m}\frac{2}{m}$$
$$(Immm)$$

may be body-centred (*I*). The symbols, in abbreviated form, are *Pmmm*, *Cmmm*, *Fmmm* and *Immm* (Fig. 8.06).

If one side of the rectangular cell is square, we choose that side as the *xy*-plane. This is the *tetragonal* lattice and its symmetries are $\frac{4}{m}\frac{2}{m}\frac{2}{m}$, often abbreviated to 4/*mmm*. We do not have it centred on the base plane, as that would merely produce a lattice of smaller squares; nor on one of the rectangular faces, since one without the other would destroy the tetrad symmetry, and both would imply a centre on the base plane. So we have only *P*4/*mmm* and *I*4/*mmm* (Fig. 8.07).

If all the faces of the unit cell are square we have the *cubic* lattice, and this has all the symmetries of the cube, denoted by $\frac{4}{m}\bar{3}\frac{2}{m}$, often abbreviated to *m3m*. It may be centred on all faces, or body-centred. Thus we have *Pm3m*, *Fm3m* and *Im3m* (Fig. 8.08).

It remains to consider the hexagonal net as base. If the third translation (and the *z*-axis) is perpendicular to the plane of the net we have a lattice, called *hexagonal*, with symmetries $\frac{6}{m}\frac{2}{m}\frac{2}{m}$, often abbreviated to 6/*mmm* (Fig. 8.09). This is the only way to preserve the hexad symmetry, but it is possible for the lattice to have triad symmetry if the third translation brings points of the net to points in the next layer opposite the centres of the triangles (Fig. 8.10). For this second arrangement the nets in successive layers are shown in Fig. 8.11(*a*).

Fig. 8.07

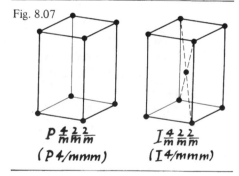

$$P\frac{4}{m}\frac{2}{m}\frac{2}{m}$$
$$(P4/mmm)$$

$$I\frac{4}{m}\frac{2}{m}\frac{2}{m}$$
$$(I4/mmm)$$

Fig. 8.08

Fig. 8.09

$$P\frac{4}{m}\bar{3}\frac{2}{m}$$
$$(Pm3m)$$

$$F\frac{4}{m}\bar{3}\frac{2}{m}$$
$$(Fm3m)$$

$$I\frac{4}{m}\bar{3}\frac{2}{m}$$
$$(Im3m)$$

$$P\frac{6}{m}\frac{2}{m}\frac{2}{m}$$
$$(P6/mmm)$$

Fig. 8.10

Fig. 8.11

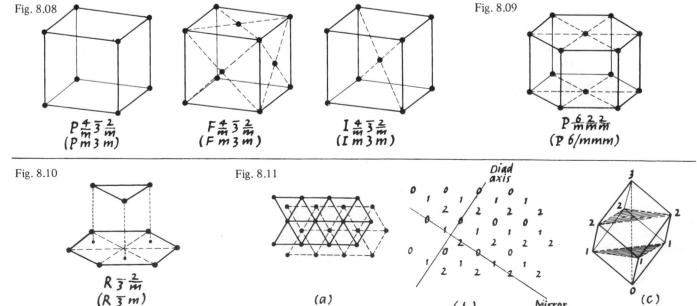

$$R\bar{3}\frac{2}{m}$$
$$(R\bar{3}m)$$

(a)

(b)

(c)

It will be noticed that a repetition of the movement from one to the other does not produce a net whose points are directly opposite those of the first one. A second repetition, however, does that. In Fig. 8.11(*b*) the numerals 0, 1, 2 indicate the positions of points in three successive layers. This is called the *rhombohedral* lattice because movements from a point in one layer to the nearest three points of the next layer could be used to define the lattice and would be the basis of a rhombohedral unit cell (Fig. 8.11(*c*)). The symmetries of this lattice are $\bar{3}\frac{2}{m}$, abbreviated to $\bar{3}m$. (A diad axis and a mirror plane are shown in Fig. 8.11(*b*). The prefix *R* is used for a rhombohedral lattice, so the complete symbol is $R\bar{3}\frac{2}{m}$ or $R\bar{3}m$ ($\bar{3}$ includes 3).

This completes the list of the 14 *Bravais lattices*, first enumerated by Auguste Bravais in 1850. They are shown all together, in diagrammatic form, in Fig. 23.14 (p. 152).

Patterns based on these lattices may have any or all of the lattice symmetries. In addition there are related patterns allowing glides and screws. This makes possible a very great variety of symmetries and in fact there are 230 types of space pattern in all.

Tins, as commonly packed (Fig. 8.12), provide an example of a tetragonal pattern, *P4/mmm*, while sugar cubes packed closely in a box form a pattern of type *Pm3m*. Matchboxes, regarded simply as rectangular blocks, and packed in the usual manner, form a *Pmmm* pattern; if, however, we consider the matches they contain (all with their heads pointing in the same direction) the pattern, though still orthorhombic, has only *P2mm* symmetry; and if, further, we consider the trays as well as the matches, the symmetry is reduced to *Pm*.

The eggs in Fig. 8.01 present quite a different arrangement. The lattice is again tetragonal, but body-centred. The full symbol for the lattice is $I\frac{4}{m}\frac{2}{m}\frac{2}{m}$, and if the eggs were spheroids the pattern would have that symmetry; but as they are narrower at one end than the other, the symmetry is only *I4mm*.

Suppose now that bricks are stacked with alternate layers arranged as in Fig. 8.13(*a*), (*b*). There is evidently a screw rotation about an axis through the points marked *O* and *O'*. (We take this as *z*-axis.) There are also half-turn axes through *O* in each of two perpendicular directions in the middle plane of a layer (the *x*- and *y*-axes) and reflexions in mirror planes in all three directions. Moreover in planes through the *z*-axis at 45° to the *x*- and *y*-axes there is glide reflexion in the *z*-direction. The symmetry is thus $P\frac{4_2}{m}\frac{2}{m}\frac{2}{c}$. (We use *c, a, b* to indicate glides in the *z*-, *x*- and *y*-directions respectively.)

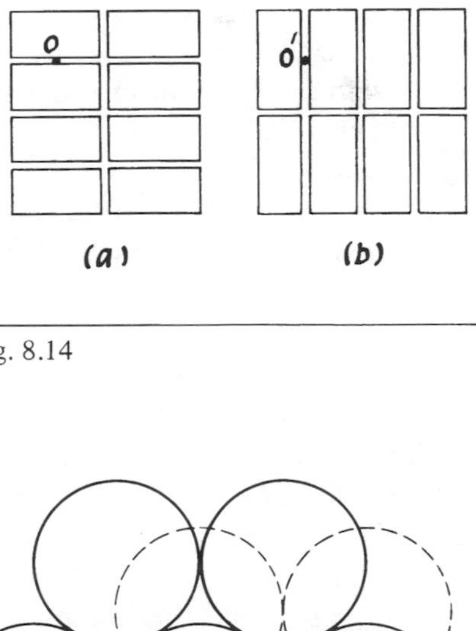

If a large number of ball bearings are shaken down in a box they tend to settle with the bottom layer arranged according to a triangular net (Fig. 8.14) and each ball of the next layer above the centre of one of the triangles (as shown by the broken lines in the figure). There are now two ways in which the third layer may arrange itself, the centre of one ball being either above *A* or above *B*. If it is above *A*, the translatory movement that takes the first layer to the position of the second cannot be repeated and is therefore not a symmetry movement of the pattern. We use instead the movement that takes the first layer to the third and the lattice is then $P\frac{6}{m}\frac{2}{m}\frac{2}{m}$; but, owing to the intermediate layer, the symmetry of the pattern is only *P3m*. If, however, the centre of a ball in the third layer is above *B*, the movement from the first layer to the second is repeated and we have a rhombohedral lattice. The symmetry of the pattern is the same as that of the lattice, namely $R\bar{3}\frac{2}{m}$.

The figures and patterns described so far are few and simple compared with

Fig. 8.12

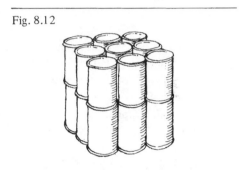

Fig. 8.13

O

O'

(a)   **(b)**

Fig. 8.14

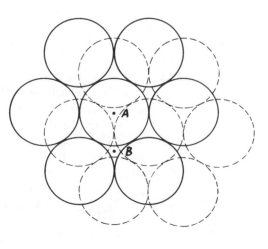

the great variety and elaboration displayed by crystals. It may well be asked why crystals have so many plane faces, inclined to one another at such a variety of angles. The crystal of iron pyrites shown in Fig. 8.15, for example, suggests the cubic form of Fig. 5.21 (p. 32) and has the same symmetry, $\frac{2}{m}\bar{3}$. But the eight corners of the cube appear to have been sliced off, in the manner of Fig. 5.18, and there are also twelve quadrilateral faces parallel to the twelve edges of the cube. In more complicated forms of the same crystal there could be faces oblique to all the edges. The reason for all this must lie in the internal structure of the crystal, i.e. in the way the molecules are packed together. The molecules are not solid objects that can be tightly packed, but the cubic form of the crystal suggests strongly that they are arranged according to a cubic lattice.

The planes of the crystal faces are in fact related to the lattice in a simple way. In Fig. 8.15 the shaded rectangles are parallel to the faces of the cube and, with the usual axes of x, y, z through the centre of the cube, their equations are $x =$ const, $y =$ const and $z =$ const. Using the coefficients of x, y, z as indices, crystallographers describe these planes by $(1, 0, 0)$, $(0, 1, 0)$ and $(0, 0, 1)$ respectively.* Of the eight triangular faces, the one in the positive octant has the equation $x + y + z =$ const (Fig. 8.16(a)) and its indices are therefore $(1, 1, 1)$. The set of eight have indices $(\pm 1, \pm 1, \pm 1)$ and are described as of *the form* $\{1, 1, 1\}$. The twelve quadrilateral faces have equations of the type $2x + y =$ const (Fig. 8.16(b)), and together constitute the form $\{2, 1, 0\}$. Their separate equations are $\pm 2x \pm y =$ const, $\pm 2y \pm z =$ const and $\pm 2z \pm x$ $=$ const, the letters being changed in cyclic order because of the triad cyclic symmetry. In more complicated crystals there might be faces of the form $\{3, 2, 1\}$ (Fig. 8.16(c)).

Another form of the pyrites crystal, shown in Fig. 8.17, has only the twelve faces of the form $\{2, 1, 0\}$, the quadrilaterals now becoming pentagons, each with four equal sides and a fifth side rather longer.

During the nineteenth century, crystals were classified in this way according to their outward forms, and were divided into 32 classes corresponding to the 32 point symmetry types illustrated in Fig. 5.25 (excluding the four 'second settings'). But, as already suggested, outward form depends on internal structure. The atoms in a crystal are arranged in an ordered repetitive pattern, which may be considered as consisting of many intersecting planes. The planes of each parallel set are equidistant from each other, with the result that when X-rays are directed at the crystal they are diffracted, and a pattern is formed on a photographic plate in the path of the emerging rays. The theory of such patterns was first studied by von Laue in 1910. Soon afterwards W.H. and W.L. Bragg showed how the photographs could be used to investigate the internal structure of the crystals, determining the angles between the planes of the atoms and the interplanar distances. Thus it became possible to deduce the arrangement of atoms within the crystal. Knowledge of this arrangement was the final step in deducing the atomic structure of the substance of which the crystal was composed. This method of determining chemical structure has been of the utmost value in physical and biological science.

Fig. 8.15

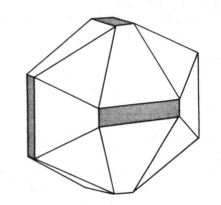

*Here the axes are in the order x, y, z.

Fig. 8.16

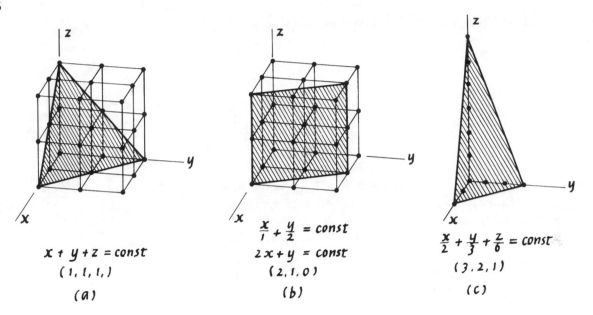

$x + y + z = const$
$(1, 1, 1,)$

(a)

$\frac{x}{1} + \frac{y}{2} = const$
$2x + y = const$
$(2, 1, 0)$

(b)

$\frac{x}{2} + \frac{y}{3} + \frac{z}{6} = const$
$(3, 2, 1)$

(c)

Fig. 8.17

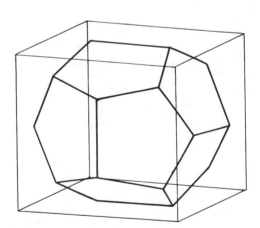

9

Patterns allowing continuous movement

We have described symmetry types in terms of rotations, reflexions, etc., and we have called these operations 'movements'. But we have been concerned with the change of position rather than the process by which it has been effected. For example, we have not distinguished between a quarter-turn clockwise and a three-quarter-turn anticlockwise. They produce the same result and so for our purposes they are equivalent. In short our 'movements' are discontinuous. They are best thought of as sudden changes of position, as shown in successive photographs of a moving object. But the changes of position can be as small as we please, and in the limit the 'movement' becomes continuous.

The patterns we have so far considered have allowed only finite 'movements', i.e. discontinuous jumps. But clearly there are some patterns that admit continuous movements. An example in three dimensions is a sheet of corrugated iron, and in two dimensions a wallpaper consisting of stripes. These allow continuous translations. Moreover a circle or a sphere allow continuous rotation.

It must be emphasized that in this book we are concerned only with static patterns. The 'movements' are imagined movements that would leave the pattern invariant. This still applies when the 'movements' are infinitesimal or continuous.

We now therefore consider patterns in one, two and three dimensions that remain invariant under continuous movements.

One dimension
In one dimension any pattern carried forward by continuous movement traces out a straight line; and a straight line has reflexion in any of its points as well as translations of any amount along its length. It is therefore impossible to have a static pattern that allows continuous translation without also allowing reflexion.

There are two kinds of one-dimensional frieze, as illustrated in Fig. 9.01, and both lead to the same limiting pattern, namely the straight line. For a symbol

Fig. 9.01

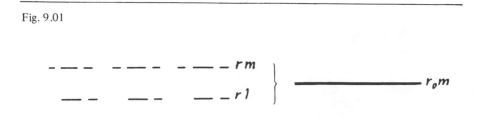

we use r_0m, the suffix zero meaning that the translation interval has been reduced to zero.

Two dimensions

In two dimensions we can have patterns allowing continuous rotation such as, for example, a hoop or a target (Fig. 9.02). This is a form of point symmetry and the symbol is ∞m, because there is reflexion in any line through the centre as well as infinitesimal rotation.

If a frieze pattern has continuous translations it reduces to continuous bands, with or without reflexion in a longitudinal mirror line (Fig. 9.03). These necessarily have transverse reflexion (i.e. in transverse mirror lines) and are limiting cases of the frieze patterns $r2mm$ and $r1m$. (The other five frieze types have the same two limiting forms.)

For a wallpaper the translation may be continuous in one or both directions. In the symbols we use p_0 and p_{00} for these possibilities. There are only three static patterns: those formed of repeated stripes, with or without longitudinal reflexion (Fig. 9.04(a), (b)); and one consisting of a uniform colour spread over the whole plane (Fig. 9.04(c)). For symbols we use p_02mm, p_01m and $p_{00}\infty m$, indicating limiting types of the wallpaper patterns $p2mm$, $p1m$ and pnm ($n \to \infty$).

Three dimensions

For point symmetry in three dimensions we can have continuous rotation, as with a round bowl, a doorknob, or any turned object; and there may also be reflexion in a plane perpendicular to the axis of rotation, as in a discus or a cotton-reel. The symbols for these types are ∞m and $\frac{\infty}{m}\frac{2}{m}$ respectively. There is also the sphere and patterns of concentric spheres, for which the symbol is $\frac{\infty}{m}\frac{\infty}{m}$.

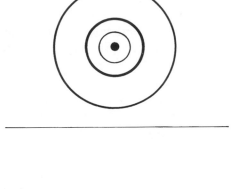

Fig. 9.02

Fig. 9.03

 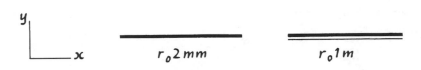

r_02mm r_01m

Fig. 9.04

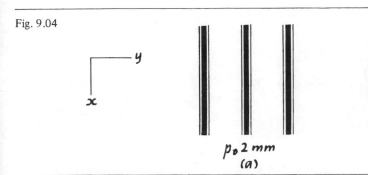 

p_02mm (a) p_01m (b) $p_0\infty m$ (c)

For line symmetry we may use any of the two-dimensional point symmetry types, namely

$$\left.\begin{array}{cccccccc} 1 & 2 & 3 & 4 & 5 & 6 & . & . & . \\ 1m & 2mm & 3m & 4mm & 5m & 6mm & . & . & . \end{array}\right\} \infty m$$

combined with continuous extension in the direction of the rotation axis (and hence reflexion in planes perpendicular to that axis). Some everyday examples are:

mouldings	$r_0 m$
metal curtain rails (Fig. 9.05)	$r_0 m2m$ or $r_0 \frac{2}{m}\frac{2}{m}\frac{2}{m}$
straight pipes or circular section	$r_0 \frac{\infty}{m}\frac{2}{m}$
uniform square posts	$r_0 \frac{4}{m}\frac{2}{m}\frac{2}{m}$
railway tunnels	$r_0 m2m$

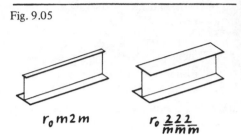

Fig. 9.05

$r_0 m2m$ $r_0 \frac{2}{m}\frac{2}{m}\frac{2}{m}$

The continuous translation may be replaced by a continuous screw movement, as illustrated, for example, by the twisted ribbon in Fig. 9.06(*a*) or the three-stranded rope in Fig. 9.06(*b*), or by an ordinary single-threaded screw. The section perpendicular to the screw has discontinuous rotational symmetry, indicated in the symbol by a subscript numeral. A superscript + or − shows whether the screw is right-handed or left-handed. Thus the three-stranded rope is $\infty_3^+ 2$ and an ordinary single-threaded screw is $\infty_1^+ 2$. (The diad rotation is demonstrated by the fact that a nut fitting a screw can be placed on it either way round.)

Alternatively, we may have continuous rotation combined with discontinuous translation. There may or may not be reflexion in transverse mirror planes and thus there are two types, $r\infty m$ and $r\frac{\infty}{m}\frac{2}{m}$, as shown in Fig. 9.07.

For plane symmetry in three dimensions the movement may be continuous in one or both directions. If it is in both, the pattern can only consist of parallel planes. There may or may not be reflexion in one such plane, as, for example, with three-ply or two-ply wood. The symbols for these types are $p_{00}\frac{\infty}{m}\frac{2}{m}$ and $p_{00}\infty m$ respectively. If there is continuous movement in one direction only, the

Fig. 9.06

$r\infty_2^+ 2$ $r\infty_3^+ 2$

(*a*)   (*b*)

Fig. 9.07

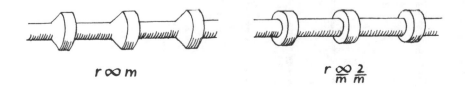

$r\infty m$ $r\frac{\infty}{m}\frac{2}{m}$

perpendicular cross-section may be of any one of the seven frieze types. Examples are:

      corrugated iron (Fig. 9.08(*a*)) $p_0 \frac{2}{b} \frac{2}{m} \frac{2_1}{m}$

      fence of overlapping slats (Fig. 9.08(*b*)) $p_0 \bar{1} \frac{2}{m}$

For space symmetry the movement may be continuous in one direction, two directions, or three non-coplanar directions, but the only example of the last is space itself. For continuity in two directions we must have parallel planes, now repeated at regular intervals. There are again two types, with or without reflexion, the cross-sections being as shown in Fig. 9.04. The symbols are $P_{00} \frac{\infty}{m} \frac{2}{m}$ and $P_{00} \infty m$. If there is continuity in one direction only, the perpendicular cross-section may be of any of the 17 wallpaper types. Everyday examples are not easily found, but a quantity of long cylinders or prisms (e.g. pencils) packed in a box gives an impression of this kind of symmetry. The symbol is $P_0 \frac{4}{m} \frac{2}{m} \frac{2}{m}$ or $P_0 \frac{6}{m} \frac{2}{m} \frac{2}{m}$, according to the method of packing.

Fig. 9.08

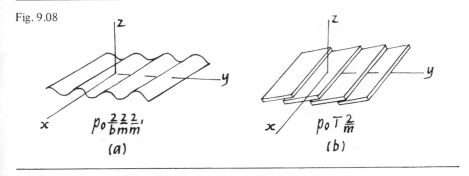

$p_0 \frac{2}{b} \frac{2}{m} \frac{2}{m}$'

(a)

$p_0 \bar{1} \frac{2}{m}$

(b)

10

Dilation symmetry

Dilation is the enlargement (or reduction) of a figure by means of lines radiating from a centre. If the centre is the point O, any point P is replaced by P', where OPP' is a straight line and $OP':OP$ is a fixed ratio. The resulting figure is geometrically similar to the original. Symmetry has been defined as the property of a figure that is self-coincident under certain transformations such as reflexion, rotation and translation. The transformations can be repeated, combined or reversed, the figure remaining invariant throughout. It might well be thought that dilation has no place here, since it is not, like the others, an isometric transformation. But repeated enlargement and repeated reduction are possible if the elements of the figure extend from the infinitely small to the infinitely large. Examples are shown in Fig. 10.01.

These are two-dimensional examples, but obviously the same principle could be applied in one dimension or in three. In one dimension there can be dilation with or without reflexion in the 'centre' point (Fig. 10.02).

In two dimensions dilation can occur with any of the symmetries of finite figures (Figs. 10.01, 10.03). In these examples the dilation and the other movements are independent, but it is also possible to have dilation linked with reflexion, in the same way that translation is so linked in a glide reflexion (Fig. 10.04). Similarly the dilation can be linked with rotation, which may be con-

Fig. 10.01

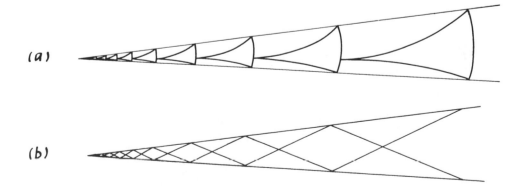

(a)

(b)

Fig. 10.02

Fig. 10.03

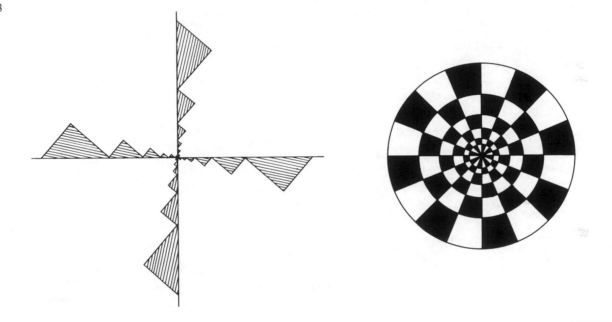

Fig. 10.04

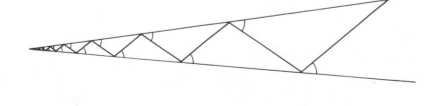

tinuous (the equiangular spiral, Fig. 10.05(*a*)) or discontinuous, as in some shells and fossils (Fig. 10.05(*b*)). In this last form there is no need for the rotation to be a sub-multiple, or even a rational fraction, of a complete turn (Fig. 10.06(*a*)). If, however, it is so, there will be simple dilations as well as the linked dilation (Figs. 10.05(*b*) and 10.06(*b*)), in much the same way as simple translations are included in a glide reflexion.

In three dimensions there are many possible types, including those in which dilation is linked with screw rotation or with inversion. Some of those with screw rotation are illustrated approximately by natural forms: shells, horns, etc., as fully described in d'Arcy Thompson's *On Growth and Form*. Inversion, in three dimensions, is an 'opposite' transformation and thus it produces a different effect from any sort of rotation. Nor is it the same as reflexion, since in fact it is reflexion linked with a half-turn. These differences could be illustrated by winding two lengths of cotton onto a cylinder or a double cone, starting from the middle and moving outwards.

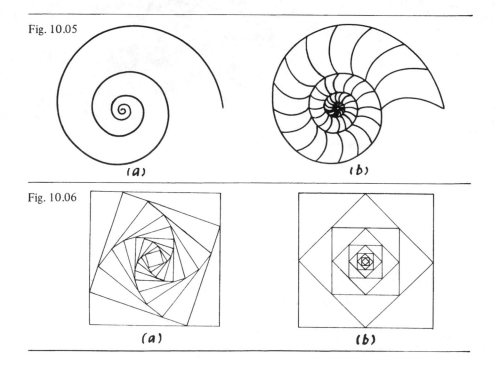

Fig. 10.05

(*a*)   (*b*)

Fig. 10.06

(*a*)   (*b*)

11

Colour symmetry

The introduction of colour greatly increases the number of symmetry types.
From the decorative point of view it is natural to think in terms of several
colours, but the simpler case of two colours is one of special importance. There
is a dichotomy of black and white, good and bad, up and down, in or out, on
or off, that runs through much of our experience. In physics we have positive
and negative charges; and a computer stores information by means of innumer-
able on–off switches.

Because of this all-pervading polarity we shall deal mainly, in this chapter,
with dichromatic symmetry. We speak in terms of colour because that at least
admits the possibility of the choice being widened; but we shall not always use
colour as a means of representation. For the dichromatic, and for polarity in
general, the most natural mathematical expression is simply + and −, though
0 and 1 can also be used, as in binary arithmetic. There are in fact several
possible representations. For example, if we wish to vary the finite pattern
$2mm$ (Fig. 11.01(a)) by having two opposite points of one polarity and two of
the other, we can show it in black and white (Fig. 11.01(b)), or by + and −
(Fig. 11.01(c)), or by introducing an extra dimension so that the figure has a
front and a back (Fig. 11.01(d) in which, following our usual convention, a
dot represents a point at the front and a ring one at the back). This last is of
course a three-dimensional pattern. It is an important form of representation
because it suggests a further development. We know that, besides the pattern
described, and the one with all four points on one side of the basic Z-plane,
there is also one with points in all four positions on both sides. The three types

Fig. 11.01

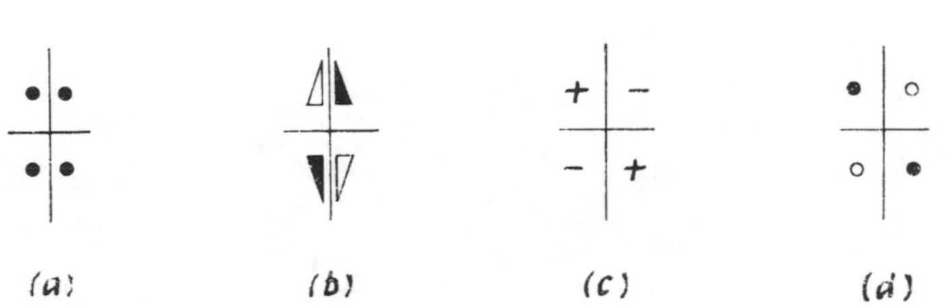

| (a) | (b) | (c) | (d) |

are shown together in Fig. 11.02. If now front and back are replaced by + and − to indicate positive and negative polarity, the (c) type becomes ±, or neutral (Fig. 11.03). In terms of colour, however, if we use white for positive and black for negative, we need an intermediate colour, grey, for the neutral type (Fig. 11.04). There are thus three kinds of dichromatic pattern, single-coloured, grey and particoloured; otherwise described as polar, neutral and of mixed polarity.

We now need a notation for dichromatic patterns. For the single-coloured types nothing new is needed, but the grey ones need to be distinguished in some way. It is, however, convenient to deal first with the particoloured types.

In a particoloured pattern the change of colour is associated with one or more of the symmetry movements and for this we use a prime attached to the symbol of the movement in question. Thus the particoloured pattern described above is denoted by $2m'm'$, indicating that there is ordinary diad rotation but that reflexion in either axis is accompanied by a change of colour. On the same principle the pattern of Fig. 11.05(a) is denoted by $2'm'm$. If a frieze is made by repeating this pattern along a row the symbol is $r2'm'm$, and for the wall-paper made by repeating it according to a rectangular net it is $p2'm'm$.

The 'grey' or neutral pattern of Fig. 11.05(b) is $2mm$ together with a change of colour associated with a complete turn. The appropriate symbol is therefore

Fig. 11.02

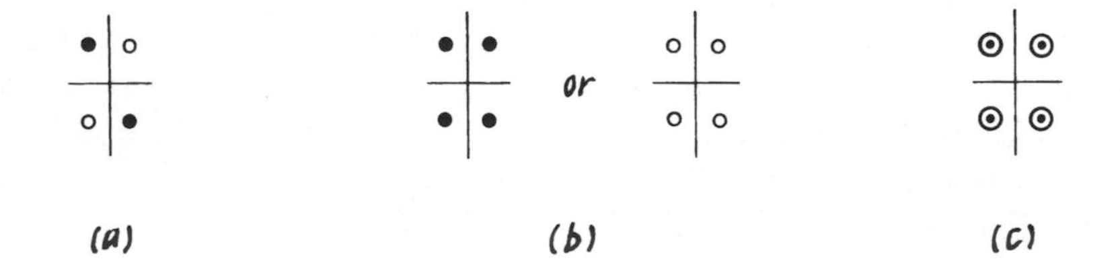

● ○	● ● or ○ ○	◉ ◉
○ ●	● ● or ○ ○	◉ ◉
(a)	(b)	(c)

Fig. 11.03

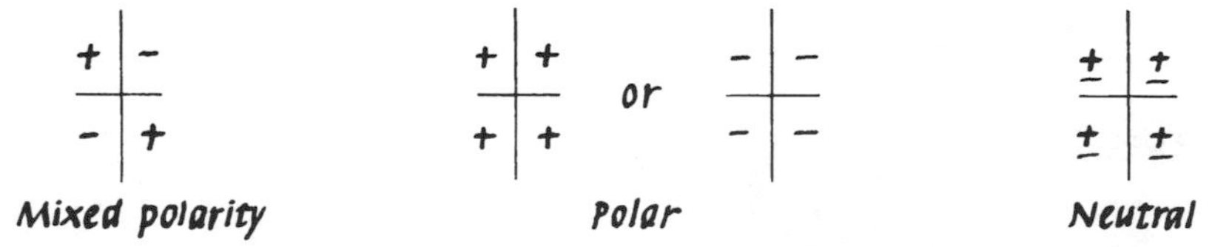

+ −	+ + or − −	± ±
− +	+ + or − −	± ±
Mixed polarity	**Polar**	**Neutral**

Fig. 11.04

Particoloured   **Single-coloured**   **'grey'**

Fig. 11.05

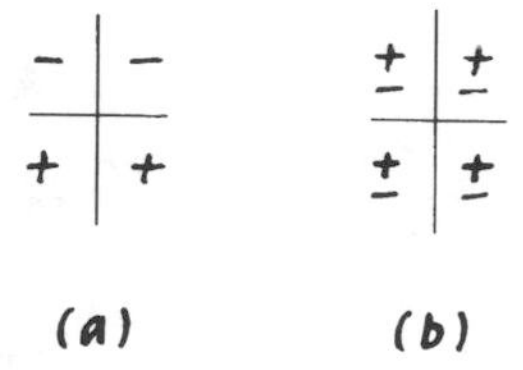

− −	± ±
+ +	± ±
(a)	(b)

1′ and this is added at the end, making 2*mm*1′. The neutral types are all designated in this way except those with odd-numbered rotation, for which 3′, 5′, . . . can be used (since a change of colour with an odd-numbered rotation automatically gives both black and white, i.e. grey, in each position).

For finite patterns there are an infinite number of types. If, however, for purposes of illustration, we restrict ourselves to the 'crystallographic' rotations 1, 2, 3, 4 and 6, there are:

10 single-coloured types	1,	2,	3,	4,	6,
	1*m*,	2*mm*,	3*m*,	4*mm*,	6*mm*;
10 'grey'	1′,	21′,	3′,	41′,	61′,
	1′*m*,	2*mm*1′,	3′*m*,	4*mm*1′,	6*mm*1′;
and 11 particoloured		2′,		4′,	6′,
		2′*mm*′,		4′*mm*′, ′	6′*mm*′,
	1*m*′,	2*m*′*m*′,	3*m*′,	4*m*′*m*′,	6*m*′*m*′.

These last are shown in Fig. 11.06.

Change of colour may also be associated with a translation. Repetition by translation is represented in one dimension by a row (*r*), in two by a net (*p* or *c*), and in three by a lattice (*P*, *A*, *B*, *C*, *F* or *I*). We now need two-coloured versions of all these, but as this chapter is mainly concerned with decorative aspects we leave the lattices for consideration in Part II.

Fig. 11.06

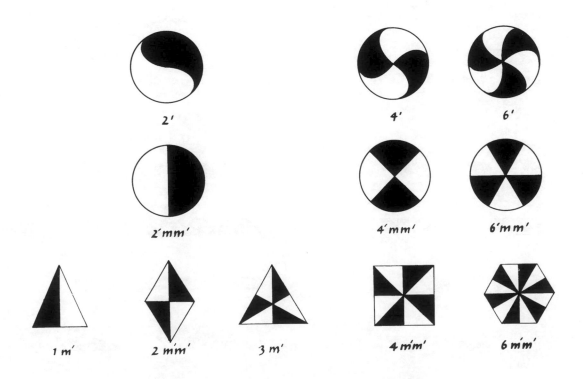

There is only one kind of two-coloured row, namely that in which the points alternate between the two colours. (In Fig. 11.07 the rings and dots represent white and black points respectively.) Such a row, or the kind of repetition that it represents, will be denoted by r'. Some examples of friezes, with their appropriate symbols, are shown in Fig. 11.08.

To determine the symbol for a given frieze it is best to mark off a unit of the pattern (the smallest part whose repetition gives the whole) and to consider first whether it contains a change of colour with simple translation. If so, the prefix is r'; if not, the colour change is associated with a rotation or a reflexion or a glide and is indicated by $2'$, m' or g'.

There are of course many friezes, such as those of Fig. 11.09, in which it is natural to ignore the background and to consider the colouring only of the more prominent parts of the design. Alternatively it is possible to partition the background into sub-cells so that we have, for example, an alternation not merely between a white dog and a black dog but between a white dog on a white ground and a black dog on a white ground. Thus we have a change of *colour scheme* rather than merely a change of colour. It is simply a question of whether we regard the background as a vacuum or whether we wish it to be included as part of the design. We can either have an alternation between, say, red posts and black posts or between red posts against a blue sky and black

Fig. 11.07

● ○ ● ○ ● ○

Fig. 11.08

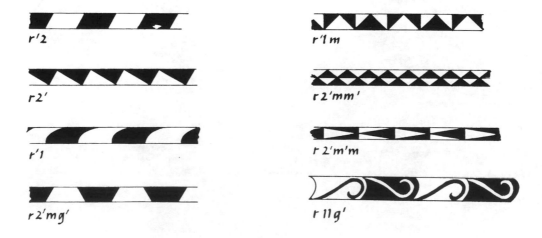

posts against a blue sky. In the latter case blue happens to occur in both colour schemes.

There are altogether 31 types of dichromatic frieze pattern, of which 7 are single-coloured, 7 'grey' and 17 particoloured. They are listed in Chapter 27.

In two dimensions there are five particoloured nets as well as the five single-coloured ones. The new ones are shown in Fig. 11.10. The direction in which the change of colours occurs is indicated where necessary by a suffix, the letters b and n being used, as before, for the directions of the y-axis and the diagonal. To see that there are only five of these nets, note first that the alternation of colour can be along one side of the cell, or both sides, or in the diagonal direction. With the rectangular cells we have all three types, and it is seen that alternation along both sides leads to a centred cell. With a square cell we cannot have alternation along one side without the other, and a centred square cell is merely a smaller one turned through 45°, still of type p'_n. With a parallelogram nothing new is obtained by using the diagonal or both sides.

Fig. 11.09

Fig. 11.10

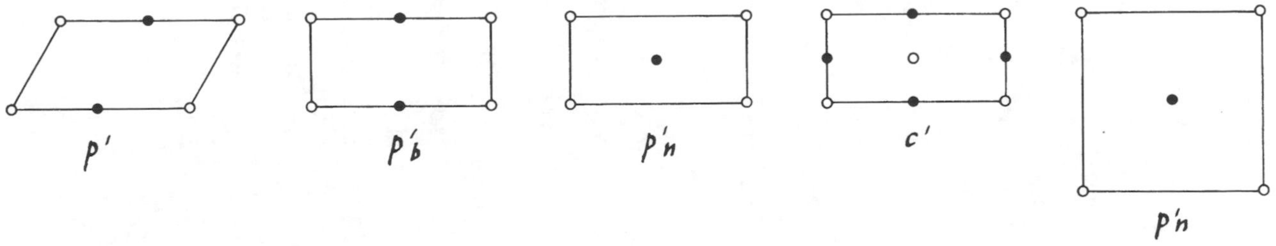

Fig. 11.11

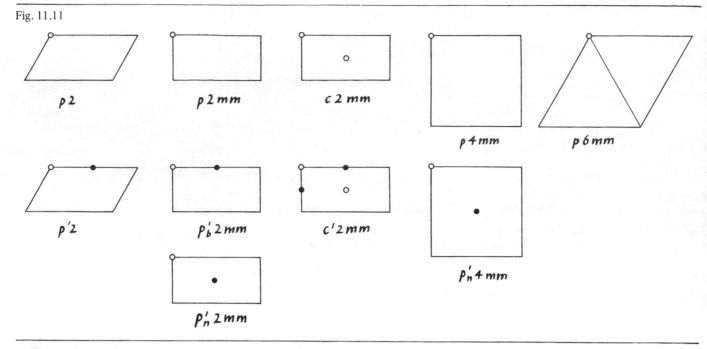

p 2   p 2 mm   c 2 mm   p 4 mm   p 6 mm

p' 2 p'_b 2 mm   c' 2 mm p'_n 4 mm

p'_n 2 mm

This makes ten nets in all, as shown together in Fig. 11.11. Here we show for each net a unit cell with only those points that 'belong' to it (in the sense that the repetition of them is just sufficient to make up the whole net). This has the advantage that in the black-and-white nets the black points are seen to be equal in number to the white, in fact forming the same pattern. We also give the symmetry symbol for each net.

Some dichromatic patterns based on these nets are shown in Fig. 11.12. (It is assumed that the background is ignored.) The explanation of the symbols is as follows.

Fig. 11.12

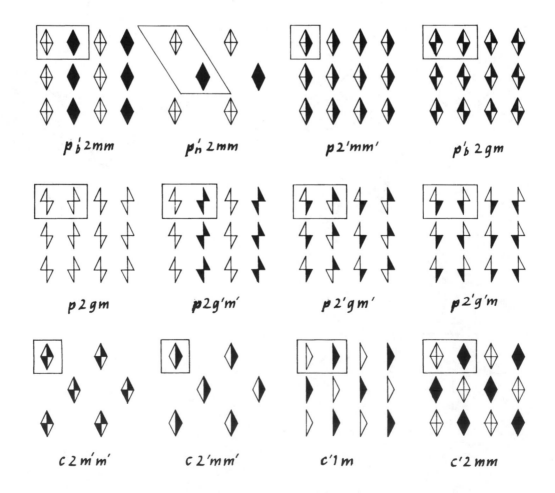

p'_b 2mm p'_n 2mm   p 2'mm' p'_b 2gm

p 2gm   p 2g'm'   p 2'gm'   p 2'g'm

c 2 m'm'   c 2'mm'   c' 1 m   c' 2 mm

The piece of each pattern enclosed in a box is a unit of the pattern, i.e. the smallest piece whose repetition gives the whole pattern. A glance at this shows at once whether there is change of colour with a simple translation. In the twelve examples of Fig. 11.12 this occurs in nos., 1, 2, 4, 11, 12, and these are the ones for which the prefix is primed. The remainder of the symbol is written without primes, though in no. 4, g could be replaced by m', and similarly in no. 1 the first m could be replaced by g'. (This occurs whenever there is a reflexion or glide line in the direction of a colour change with translation. It is in fact the only way in which a coloured reflexion line and an uncoloured glide line, or vice versa, can coincide.)

If, however, there is no change of colour with simple translation, the prefix is p or c (unprimed) and the alternation of colour is indicated in the remaining parts of the symbol. We thus distinguish two types of dichromatic pattern.

We have not given a lower line in the symbols to show the reflexions or glides in intermediate lines. The reason for this is that, like the numeral for the rotation, they follow automatically, but unlike the numeral they do not add much to the characterization of the symmetry type. It may, however, be noticed that, where there is alternation of colour in the direction of the y-axis (p'_b), there are four reflexion or glide lines per unit, at quarter-unit intervals,

Fig. 11.13

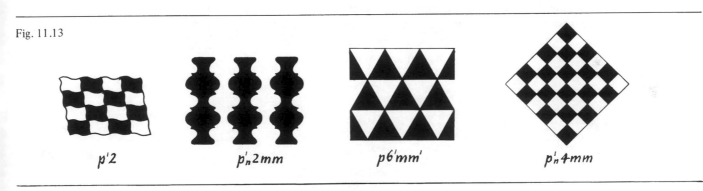

$p'2$ $p'_n 2mm$ $p6'mm'$ $p'_n 4mm$

Fig. 11.14

one pair coloured and one uncoloured (m and m'). With p'_n this happens for both the x- and y-directions, the pairs being m and g' or m' and g. With the net c' it does not arise, as a unit shift is only half the diagonal of the centred cell.

In Fig. 11.13 we show some dichromatic mosaics. These are patterns covering the whole plane, so the question of background does not arise. The use of various types of symmetry in the design of mosaic patterns has been fully explored by M.C. Escher.* We show examples of his work in Figs. 11.14 and 11.15 (see also Fig. 0.05, p. 4). The pattern of Fig. 0.05 is of type $p4$, while that of Fig. 11.14 is dichromatic, with glide reflexions, symbol $p2'gg'$. Fig. 11.15 shows a three-coloured pattern of type $p6^{(3)}$ (see below).

There are altogether 80 types of dichromatic pattern in the plane. Of these, 17 are single-coloured (black or white), 17 grey and 46 particoloured. They correspond to the 80 types of plane ornament in three dimensions. The particoloured ones are illustrated in Fig. 27.09 (p. 200).

Dichromatic patterns in three dimensions are of less interest from the decorative point of view. They will be discussed briefly in Part II, but we note here that the 230 types of uncoloured space pattern give 1651 dichromatic varieties.

We have illustrated the particoloured types in black and white, but of course any two colours will do, and indeed any two colour schemes. Many commercial friezes and wallpapers show an alternation of colour schemes and are counted, for our purposes, as dichromatic. When we speak, in the following sections, of polychromatic patterns we mean those that show not an alternation but a sequence of three or more colours or colour schemes.

*See C.H. Macgillavry's *Symmetry Aspects of M.C. Escher's Periodic Drawings.*

Fig. 11.15

Polychromatic symmetry

If more than two colours are used they must be in a definite sequence, the steps of the sequence being associated with a symmetry movement of the pattern. Moreover any movement of the pattern must either leave the colours unchanged or move all of them one step along the sequence, or two steps, or more. Thus if the sequence is black, red, white, pink, black, . . . , we can have such patterns as that of Fig. 11.16(a), in which translation in one direction leaves the colours unchanged but in other directions changes them by varying numbers of steps. Note that in this pattern there are no reflexions, the net being of parallelogram type. But reflexions or glides can occur in patterns based on a rectangular net, centred or otherwise (Fig. 11.16(b), (c), (d)).

For patterns of this kind any number of colours may be used, and there are thus an infinite number of types. But where rotations are concerned the possibilities are much more limited. It will be noticed that in Fig. 11.16(a), if the colours were removed, the pattern would have reflexion and diad rotation: but with the colours as shown neither of these movements is possible. Thus the introduction of colours tends to reduce the symmetry. Both diad rotation and reflexion, if repeated, bring a pattern back to its original position, so those movements can be associated with an alternation of colours, but not with a sequence of three or more. (They may, however, accompany a two-step colour change in a sequence of four as, for example, reflexion does in the frieze shown in Fig. 11.17(c).)

Fig. 11.16

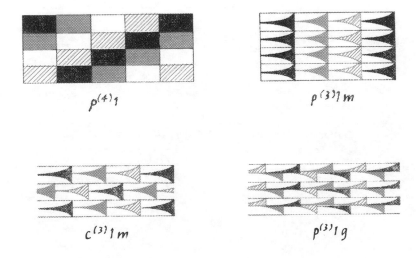

$p^{(4)}1$

$p^{(3)}1m$

$c^{(3)}1m$

$p^{(3)}1g$

Multicoloured frieze patterns can be based only on the friezes $r1$, $r11m$ and $r11g$. Examples are shown in Fig. 11.17. For symbols we indicate the number of colours as a superscript in parentheses, for example $r^{(4)}$ for change of colour with translation, $g^{(4)}$ for change of colour with glide reflexion. (For patterns of two colours we used r', equivalent to $r^{(2)}$. For a two-colour change in a four-colour sequence the $r^{(2)}$ form is appropriate.) It is to be noted that the two friezes in Fig. 11.17(a) are of the same type, though derived from different uncoloured types, $r1$ and $r1m$.

Multicoloured wallpaper patterns, as already mentioned, are of two kinds: those in which rotations do not occur, the colour sequence being associated with a translation, and those allowing rotations. For the first kind there are four basic types, as already shown in Fig. 11.16, and any number of colours may be used; but for the second we are limited to the 'crystallographic' numbers 3, 4 and 6. Change of colour with translation is represented by a coloured net, of one of the types shown in Fig. 11.18. (The colours are represented by numbers.) The first three are those in which any number of colours may be used. The fourth is a special case of the third, with three colours only and a hexagonal pattern. It is the only multicoloured net that allows any sort of rotation.

Fig. 11.17

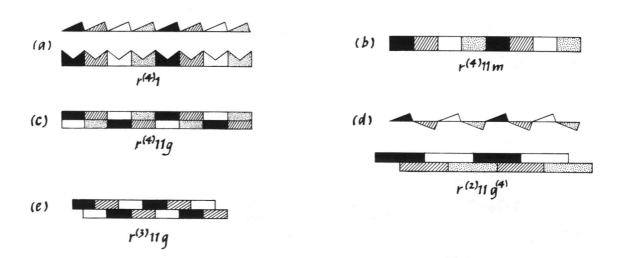

		(a) $r^{(4)}1$				(b) $r^{(4)}11m$
		(c) $r^{(4)}11g$				(d) $r^{(2)}11\,g^{(4)}$
		(e) $r^{(3)}11g$				

Fig. 11.18	Parallelogram	Rectangular	Centred rectangular	Hexagonal
	1 2 3 1 −	1 2 3 1 −	1 2 3 1 −	1 2 3 1 −
	1 2 3 1 −	1 2 3 1 −	3 1 2 3 −	3 1 2 3 −
	1 2 3 1 −	1 2 3 1 −	1 2 3 1 −	1 2 3 1 −
	$p^{(3)}$	$p^{(3)}$	3 1 2 3 −	3 1 2 3 −
			$c^{(3)}$	$p^{(3)}$
			1 2 3 4 1 −	
			3 4 1 2 3 −	
			1 2 3 4 1 −	
			3 4 1 2 3 −	
			$c^{(4)}$	

Fig. 11.19

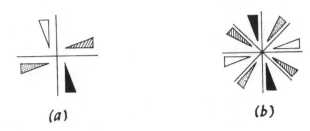

(a)   (b)

With rotation, the sequence of colours may occur once in a revolution (Fig. 11.19(*a*)) or twice (Fig. 11.19(*b*)) or more times. The symbols for these patterns are $4^{(4)}$ and $8^{(4)}$ respectively. The symbol $6^{(3)}$, like $8^{(4)}$, implies two colour sequences in a revolution, and $6^{(2)}$ implies three.

For a finite pattern there are an infinite number of types, but for a wallpaper pattern we are restricted to the 'crystallographic' rotations 2, 3, 4 and 6, and it follows that the number of colours is similarly restricted. This is a severe limitation and in fact there are only eleven polychromatic types that allow rotations.* These are shown in Fig. 11.20. One of them, $p6^{(3)}$, is also illustrated by M.C. Escher's design reproduced in Fig. 11.15 (p. 66).

*Belov, in Shubnikov & Belov's *Colored Symmetry*, designates these types by symbols for their three-dimensional representations. He gives 15 types, but only by counting both members of the four enantiomorphic pairs based on $4^{(4)}$, $3^{(3)}$, $6^{(6)}$ and $6^{(3)}$. These are mere reversals of the colour sequence.

Fig. 11.20

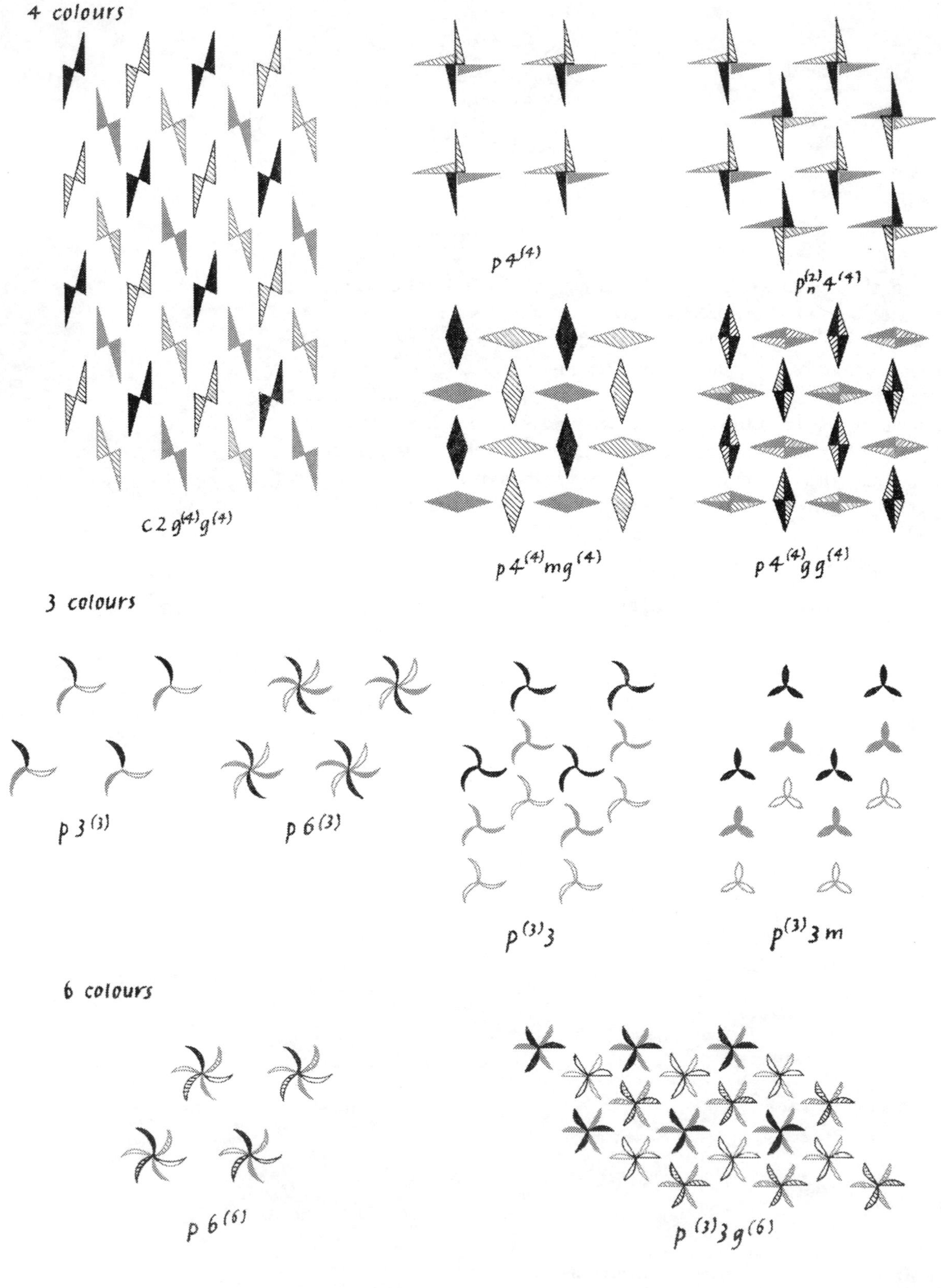

4 *colours*

$c\,2\,g^{(4)}g^{(4)}$

$p\,4^{(4)}$

$p_n^{(2)}\,4^{(4)}$

$p\,4^{(4)}mg^{(4)}$

$p\,4^{(4)}g\,g^{(4)}$

3 *colours*

$p\,3^{(3)}$

$p\,6^{(3)}$

$p^{(3)}3$

$p^{(3)}3\,m$

6 *colours*

$p\,6^{(6)}$

$p^{(3)}3\,g^{(6)}$

12

Classifying and identifying plane patterns

While crystallography and atomic science provide important fields for the application of symmetry principles in three dimensions, symmetry in the plane is more naturally associated with decorative work. Here there are two chief ways in which the ideas of symmetry can be used. For anyone observing a piece of decoration or construction, in art or in nature, there is an added pleasure to be obtained by noticing symmetries; and for the designer, professional or amateur, the appreciation and use of symmetry can help greatly in the making of new and interesting types of design. In the two remaining chapters of Part I we deal with these two aspects of symmetry in the plane.

Cells and sub-cells

The first step, in looking at a repeating pattern, is to identify the mode of repetition, as represented by a net of points. This is usually done by observing any one feature of the design and the points where it is repeated, checking that each of these points has the same environment. The points of the net suggest parallelograms and one of these, of minimum area, is a *cell*. The cell may be chosen in a variety of ways, but it is natural to choose one that exhibits symmetrical properties (e.g. a rectangle or a rhombus) when possible. Thus in Fig. 12.01 the net has a rhomboidal cell, indicated by dots, and in Fig. 12.02 a square cell is chosen (in either of two positions). It is nevertheless sometimes convenient to use as a unit of pattern a portion with a different boundary, not necessarily a parallelogram. It must be of the same area as a parallelogram cell and such that its repetition by translation gives the whole pattern. Examples are given in the same figures.

A single cell (or unit of pattern) can be broken down into sub-cells (or sub-units) containing identical pattern elements related not by translation but by rotation, reflexion or glide reflexion. In Fig. 12.02 there are eight triangular sub-cells (or a unit of pattern, as shown, may be divided, rather more obviously, into eight square sub-units).

Fig. 12.01

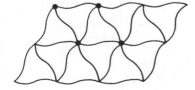

Cell

Alternative
units of pattern

Fig. 12.02

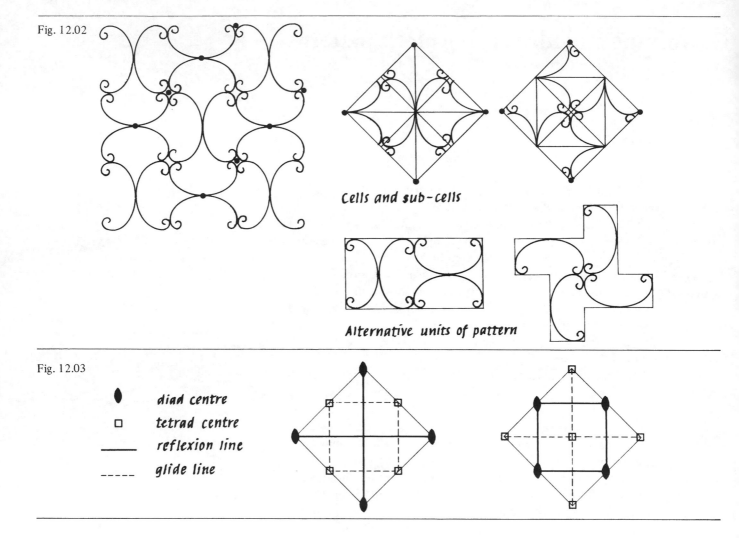

Cells and sub-cells

Alternative units of pattern

Fig. 12.03

●   diad centre
□   tetrad centre
——   reflexion line
‑ ‑ ‑   glide line

For a given pattern the cell may be varied in position, as shown in Fig. 12.02. If it is to break up into sub-cells its vertices should be related to the rotation centres and reflexion lines, not primarily to glide lines. The symmetry charts of Fig. 12.03 correspond to the two cells shown in Fig. 12.02.

In Fig. 12.04 we show cells and sub-cells for each of the 17 'wallpaper' types. For those in which there is triad symmetry the parallelogram cell cannot be divided into three sub-cells, so we use instead a hexagonal pattern-unit. This is convenient also when the symmetry is hexad.

Coloured sub-cells

In two-coloured patterns the basic cell can suffer colour reversal as well as reflexion and can thus now appear in four different forms. These are illustrated in Fig. 12.05, where the numerals 1 and 2 refer to the colours (or colour schemes) and reflexion is indicated by the form of the half-arrows. It must again be emphasized that the word 'colour' may always be replaced by 'colour scheme'. Thus colour 1 might refer to a design in red and green, and colour 2 the same design with colours reversed (or in two quite different colours).

Bearing in mind that the sub-cells must be equally divided between the two colours, it is easy to derive coloured types from those of Fig. 12.04. Thus with

Fig. 12.04

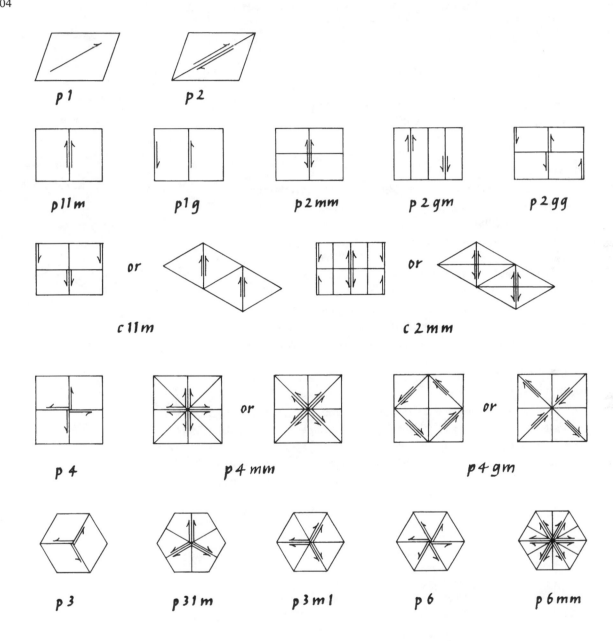

p1   p2

p11m   p1g   p2mm   p2gm   p2gg

or

c11m   c2mm

or   or

p4   p4mm   p4gm

p3   p31m   p3m1   p6   p6mm

Fig. 12.05

Fig. 12.06

$p4g'm'$ $p4'gm'$ $p4'g'm$

Fig. 12.07

$p4gm$, for example, there are three ways of dividing the eight sub-cells, as shown in Fig. 12.06. (It will be recalled that the three movements represented by 4, g and m are interdependent. That is why we cannot have change of colour associated with only one of them.)

There may also be change of colour with translation. The diagrams are drawn by adding elements of the second colour according to one of the five coloured nets shown in Fig. 11.11 (p. 64). For $p4gm$ this gives 16 sub-cells (Fig. 12.07).

A table showing all the coloured types will be found at the end of this chapter (Table 12.1).

p'_n4gm

Multicoloured patterns

With two colours we can always use the same cell as for the corresponding single-coloured pattern. Thus the diagrams of Figs. 12.06 and 12.07 are derived from that for $p4gm$ (Fig. 12.08(a)). But with more than two colours this is not always possible. When there is a glide reflexion in four colours a cell of double area is needed, as, for example, in $p4^{(4)}g^{(4)}m$ (Fig. 12.08(b)). The same applies to $c2g^{(4)}g^{(4)}$ (Fig. 12.09) and for this we show two complete cells, because the glides in two perpendicular directions lead to a centred net and it is therefore desirable to show the relative positions of the cells. With $p4^{(4)}g^{(4)}m$ there is again a centred net, this time a square one. But a square centred net gives new squares at 45° to the original ones, so it is convenient to turn the cell diagram through that angle and to rearrange its boundaries as a square (Fig. 12.08(c)). The symbol is then $p4^{(4)}mg^{(4)}$. Similar considerations apply to $p4^{(4)}g^{(4)}g$, which becomes $p4^{(4)}gg^{(4)}$.

With three or six colours we may need a cell of area three times that for the single-coloured pattern. This happens with $p^{(3)}3$, $p^{(3)}3m$ and $p^{(3)}3g^{(6)}$, and for each of these we again show two complete cells.

In Fig. 12.09 we include cells and sub-cells for each of the eleven types in which rotation occurs. The number of colours can only be 3, 4 or 6. Patterns in

Fig. 12.08

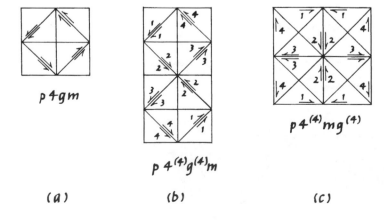

$p4gm$ $p4^{(4)}g^{(4)}m$ $p4^{(4)}mg^{(4)}$

(a) (b) (c)

Fig. 12.09

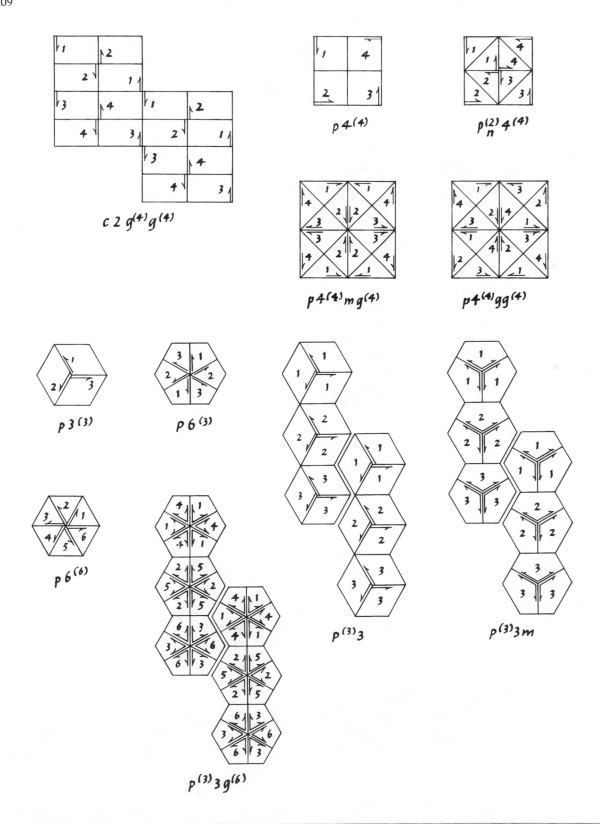

$c\,2\,g^{(4)}g^{(4)}$

$p\,4^{(4)}$

$p_n^{(2)}\,4^{(4)}$

$p\,4^{(4)}m\,g^{(4)}$

$p\,4^{(4)}g\,g^{(4)}$

$p\,3^{(3)}$

$p\,6^{(3)}$

$p\,6^{(6)}$

$p^{(3)}\,3$

$p^{(3)}\,3m$

$p^{(3)}\,3\,g^{(6)}$

which there are no rotations can be made with any number of colours. Change of colour can accompany translation according to a coloured net of one of the types shown in Fig. 11.16 (p. 67). With the rectangular or centred rectangular nets there can be mirror lines or glide lines, but in one direction only. There are four basic types, $p^{(n)}1$, $p^{(n)}1m$, $p^{(n)}1g$ and $c^{(n)}1m$. Cells and sub-cells for these, with n equal to 3, are shown in Fig. 12.10(a). For a glide reflexion the glide distance is necessarily half the (uncoloured) translation distance. If there are both reflexions and glides the net is a centred one, as is shown more clearly in the brickwork pattern of Fig. 12.10(b).

Pattern identification

Facility in identification comes only with practice, but for this there is ample opportunity in everyday life, as we are constantly faced with symmetrical objects and patterns.

Finite plane patterns present little problem, as it is only necessary to observe the centre of rotation or pole and count the number of repetitions (n) of the basic pattern in a full rotation; then to notice whether there are reflexion lines (m) through the pole. Either there are no reflexion lines and the pattern is of type n ($n = 1, 2, 3, \ldots$) or there are n such lines and the pattern is nm.

Frieze patterns are almost equally easy to identify. The first step is to observe the interval of repetition and use it to mark off a cell. The boundaries of the cells are conveniently taken at points where there are reflexion lines or rotation centres. In Fig. 12.11, for example, the boundaries may be chosen in either of these ways. The cell then breaks up into four sub-cells, related to one another by a half-turn or by transverse reflexion or by a glide. The pattern is thus of type $r2mg$. In this sort of way it is easy to determine which of the seven types any frieze belongs to.

'Wallpaper' patterns are rather more difficult. Cells and sub-cells are identified in the way already described. Some rotations and reflexions are immedi-

Fig. 12.10

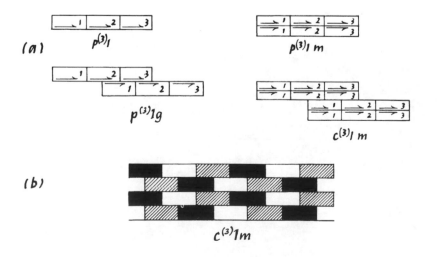

(a) $p^{(3)}1$ $p^{(3)}1m$

$p^{(3)}1g$ $c^{(3)}1m$

(b)

$c^{(3)}1m$

Fig. 12.11

Fig. 12.13

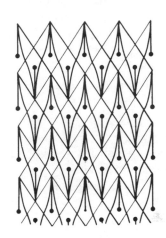

ately apparent, but a further search for these and for glide reflexions must be made, bearing in mind the following guides:

1. Parallel reflexion lines occur at repetition intervals, and midway between any neighbouring pair of such lines is an intermediate reflexion or glide line. The same applies to parallel glide lines.

2. Intersecting reflexion or glide lines imply a rotation. For two reflexion lines the centre of rotation is at the intersection point, but if one or both of the lines are glide lines it is elsewhere. Conversely, a rotation and a reflexion or glide imply another reflexion or glide.

3. Rotation centres at repetition distance imply further centres of the same (and sometimes lower) multiplicity, according to the diagrams in Fig. 12.12.

The net, the principal rotations, and the reflexion and glide lines are now determined and invariably form one of the combinations listed in Table 4.2 (p. 16) and illustrated in Fig. 4.36 (p. 25). Thus the pattern is identified.

In Fig. 12.13, for example, it is easy to identify the rectangular net (primitive, not centred), the vertical reflexion lines and the horizontal glide lines. This implies diad rotation, and diad centres are seen at the mid-points of the sides of the rhombuses. The pattern is therefore of type *p2gm*.

The arabesque shown in Fig. 12.14 is much more difficult. Two points with identical environments are marked in the figure and evidently the net is a square one, tilted at 45°. A vertical reflexion line is marked, the neighbouring parallel lines being at the sides of the diagram. An intermediate glide line is shown. The points of the net are seen to be centres of tetrad rotation and this implies that there should be further reflexion or glide lines at 45° to the ones already observed. In fact they are glide lines (as must occur when the reflexion lines do not pass through the tetrad centres). The pattern is thus of type *p4gm*.

Two-coloured patterns have been considered in Chapter 11. The cell, or unit of pattern (e.g. Fig. 11.12, p. 64), may or may not show a change of colour with translation. If it does, the pattern is based on one of the particoloured

Fig. 12.14

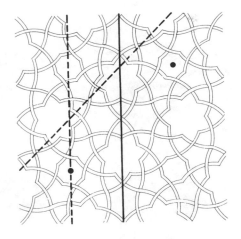

Fig. 12.12

● diad centre

▲ triad

◆ tetrad

⬣ hexad

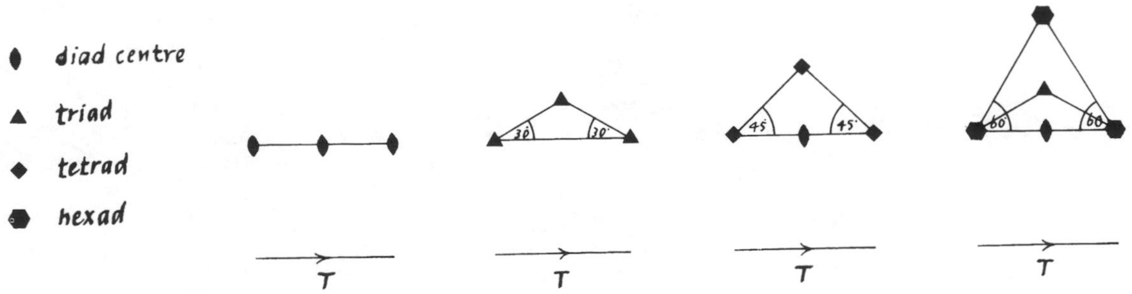

nets shown in Fig. 11.11 and it is not then necessary to associate colour change with the other movements. A primed prefix is used for these patterns. If, however, there is no change of colour with translation the prefix is a plain p or c and we look for colour change in the movements relating the sub-cells to one another. Thus the design of Fig. 12.15 is of type $p1g'$. Further examples are shown in Fig. 11.13 (p. 65) and, for frieze patterns, in Fig. 11.08 (p. 62).

With more than two colours the same principles apply. If the colour changes with translation in one direction only (this includes frieze patterns) we have only the four basic types shown in Fig. 12.10(*a*). Other possibilities are limited to the eleven types of Fig. 11.20 (p. 70) and this makes identification comparatively easy.

By way of summary we give a table showing all the types of coloured repeating pattern in two dimensions (Table 12.1). Finally we give some patterns for practice in identification (Fig. 12.16).

Fig. 12.15

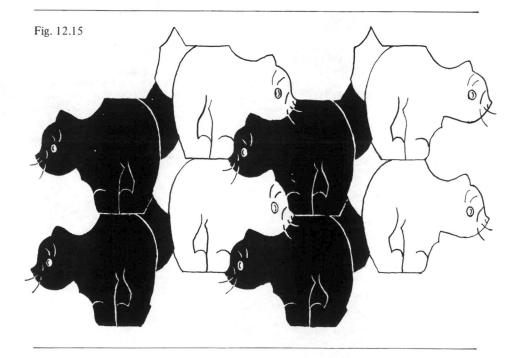

Table 12.1

Basic type	2 colours		3, 4 or 6 colours		n colours $(n \geqslant 3)$
	Plain row or net with coloured cells	Coloured row or net with plain cells	Plain with coloured cells	Coloured, with plain or coloured cells	Coloured, with plain cells
Frieze patterns					
$r1$	—	$r'1$	—	—	$r^{(n)}1$
$r2$	$r2'$	$r'2$	—	—	—
$r1m$	$r1m'$	$r'1m$	—	—	—
$r11m$	$r11m'$	$r'11m$	—	—	$r^{(n)}11m$
$r2mm$	$r2m'm', r2'm'm, r2'mm'$	$r'2mm$	—	—	—
$r11g$	$r11g'$	$r'11g$	—	—	$r^{(n)}11g$
$r2mg$	$r2m'g', r2'm'g, r2'mg'$	$r'2mg$	—	—	—
Wallpaper patterns					
$p1$	—	$p'1$	—	—	$p^{(n)}1$
$p2$	$p2'$	$p'2$	—	—	—
$p1m$	$p1m'$	p'_b1m, p'_b11m, p'_n1m			$p^{(n)}1m$
$p1g$	$p1g'$	p'_b1g, p'_b11g, p'_n1g	—	—	$p^{(n)}1g$
$c1m$	$c1m'$	$c'1m$	—	—	$c^{(n)}1m$
$p2mm$	$p2m'm', p2'm'm$	p'_b2mm, p'_n2mm	—	—	—
$p2mg$	$p2m'g', p2'm'g, p2'mg'$	$p'_b2mg, p'_b2gm, p'_n2mg$	—	—	—
$p2gg$	$p2g'g', p2'gg'$	p'_b2gg, p'_n2gg	—	—	—
$c2mm$	$c2m'm', c2'mm'$	$c'2mm$	$c2g^{(4)}g^{(4)}$	—	—
$p4$	$p4'$	p'_n4	$p4^{(4)}$	$p_n^{(2)}4^{(4)}$	—
$p4mm$	$p4m'm', p4'mm', p4'm'm$	p'_n4mm	—	—	—
$p4gm$	$p4g'm', p4'gm', p4'g'm$	p'_n4gm	$p4^{(4)}mg^{(4)}, p4^{(4)}gg^{(4)}$	—	—
$p3$	—	—	$p3^{(3)}$	$p^{(3)}3$	—
$p3m1$	$p3m'1$	—	—	$p^{(3)}3m1$	—
$p31m$	$p31m'$	—	—	—	—
$p6$	$p6'$	—	$p6^{(6)}, p6^{(3)}$	—	—
$p6mm$	$p6m'm', p6m'm, p6'mm'$	—	—	$p^{(3)}3g^{(6)}$	—

Fig. 12.16   **PATTERNS FOR IDENTIFICATION**

Finite patterns

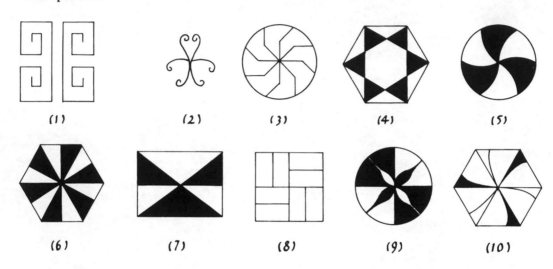

Frieze patterns

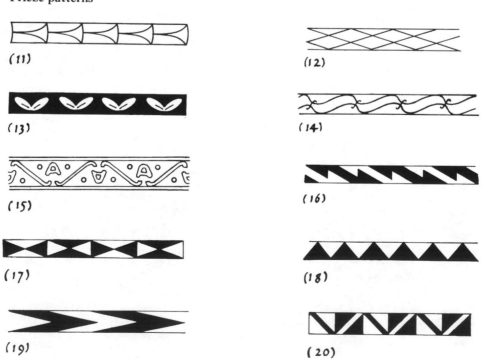

Wallpaper patterns

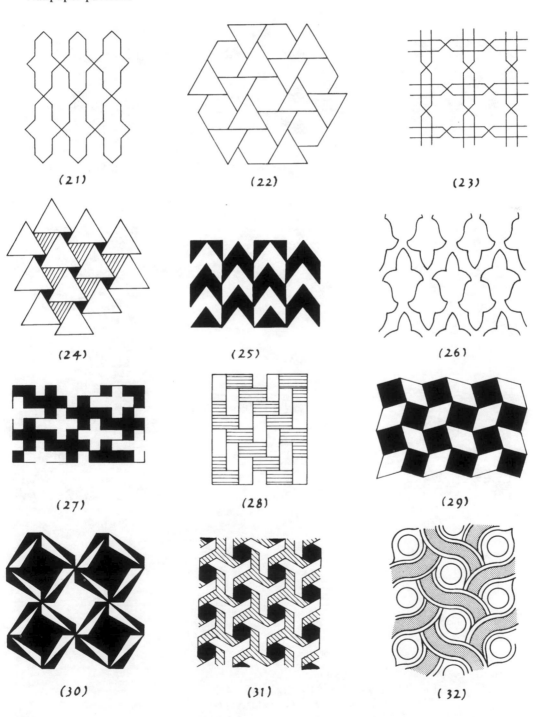

(21)   (22)   (23)

(24)   (25)   (26)

(27)   (28)   (29)

(30)   (31)   (32)

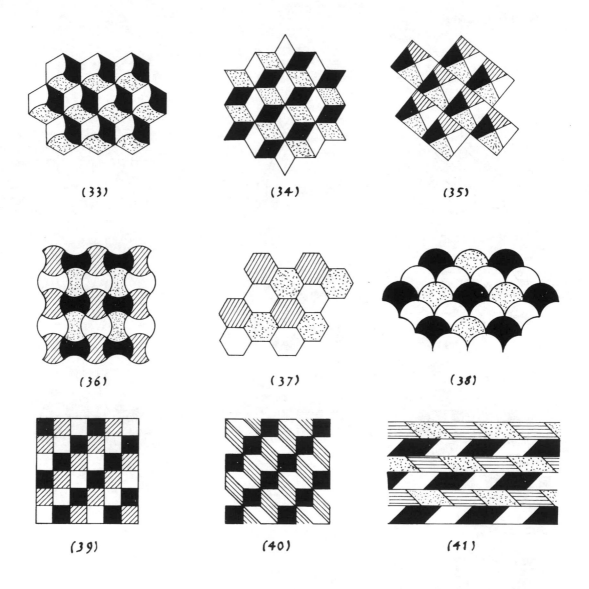

(33)   (34)   (35)

(36)   (37)   (38)

(39)   (40)   (41)

ANSWERS

(1) 2*mm*. (2) 1*m*. (3) 8. (4) 6*mm*. (5) 6'. (6) 6*m'm'*. (7) 2'*m'm*. (8) 4.
(9) 4'*m'm*. (10) 6'. (11) *r*11*m*. (12) *r*2*mg*. (13) *r*1*m*. (14) *r*2. (15) *r*2*mg*.
(16) *r*'2. (17) *r*'2*mm*. (18) *r*2'*mg*'. (19) *r*'11*m*. (20) *r*2*m'g*'. (21) *c*2*mm*.
(22) *p*6. (23) *p*4*mm*. (24) *p*3. (25) *c*'1*m*. (26) *p*2*gm*. (27) *p*'2. (28) *p*2*g'g*'.
(29) *p*4*gm*. (30) p'_n4. (31) *p*6'. (32) *p*2*gg*, (33) $p3^{(3)}$. (34) $p6^{(3)}$.
(35) $p_n^{(2)}4^{(4)}$. (36) $p4^{(4)}mg^{(4)}$. (37) $p^{(3)}3m$. (38) *c*'1*m*. (39) $c^{(3)}1m$.
(40) *c*2*m'm*'. (41) $c2g^{(4)}g^{(4)}$.

13

Making patterns

There are a number of ways in which a knowledge of the symmetry properties of patterns can be used to derive new designs. Some of these are almost automatic, others more under the designer's control. We may start by taking a basic cell and dividing it into sub-cells, in each of which we put the same pattern. In doing this we note that the cell-lines will normally be eliminated in the final complete pattern, which may itself cross the cell boundaries. For example the sub-cell pattern of Fig. 13.01(a) gives the completed design shown alongside, using the sub-cell arrangement for p4gm (Fig. 12.04, p. 73).

It will be noticed that the two curves bounding the shaded area in the sub-cell are arranged so as to form continuous lines crossing the sub-cell boundaries in the completed design. This is shown in the double sub-cell of Fig. 13.01(c), and indeed the pattern can be conveniently drawn by using the single curve of that figure and applying to it the rotations and reflexions of p4mg.

Lines crossing the sub-cell boundaries are of course a common feature of many designs. Viewed in this way the underlying simplicity of some apparently elaborate patterns is quite astonishing. For example the beautiful arabesque of Fig. 13.02 is constructed by repetition of a single irregular line crossing five sub-cells. It is possible, though not entirely easy, to obtain many other patterns by applying the same principle with other shapes of line and other arrange-

Fig. 13.02

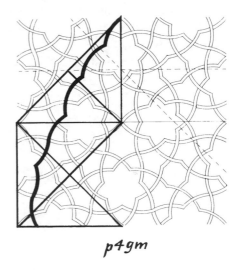

p4gm

Fig. 13.01

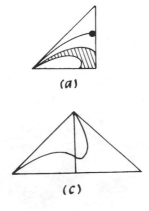

(a)

(c)

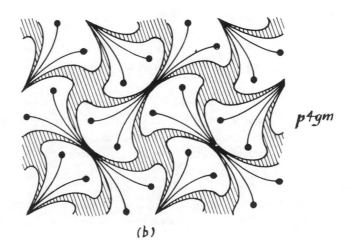

p4gm

(b)

ments of sub-cells. It is best to apply the rotations and reflexions to the whole line-segment, using tracing paper or a template if necessary.

An alternative approach to pattern construction is to mark two or three points in one of the sub-cells, and corresponding or related points in each of the others, and to join these points without intersections in some regular way, thus forming a network which can be used as basis for a pattern. An example of this is shown in Fig. 13.03.*

A third method is to distort the boundaries of the sub-cells in such a way that the plane is filled with similar interlocking shapes, as, for example, in Fig. 13.04. Such a pattern can easily be coloured systematically to produce one of the coloured symmetry types, as shown here.

The drawing of patterns by any of these methods is greatly simplified if sheets are available marked with the five standard nets (parallelogram, rectangle, rhombus, square and hexagon) sub-divided into triangles as sub-cells. An even greater simplification, although it does introduce some restriction on the variety of patterns produced, is to have as a grid a suitable basic pattern from which many designs can be produced by using only certain of the background lines and by colouring areas between the lines. The grid can be relatively simple, like that of Fig. 13.05, or more elaborate, as in the well-known 'Altair Designs'

Fig. 13.03

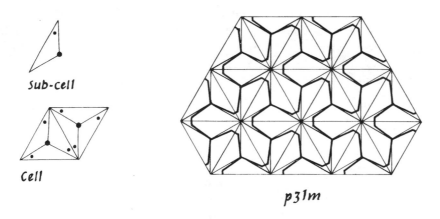

Sub-cell

Cell

p31m

(Fig. 13.06) produced for use in schools. (Note the variety of polygons that occur in the latter.) Either of these grids can be used as a basis for any wall-paper type except those having triad or hexad rotation. Marked on Fig. 13.05 is an indication of how a pattern of type *p4gm* can be obtained, even though the basic grid is of form *p4mm*.

Clearly another method of using a basic grid is to replace the straight lines by curved ones conforming to the desired symmetry group, the method used

Fig. 13.04

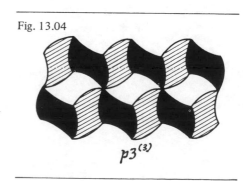

p3(3)

*A large number of examples of patterns derived in this way will be found in Schubnikov & Koptsik: *Symmetry in Science and Art*, pp. 176–9.

Fig. 13.05

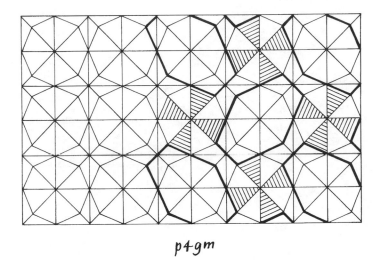

p4gm

Fig. 13.06 'Altair Design' developed by Dr Ensor Holiday (Published by Longman in 1973)

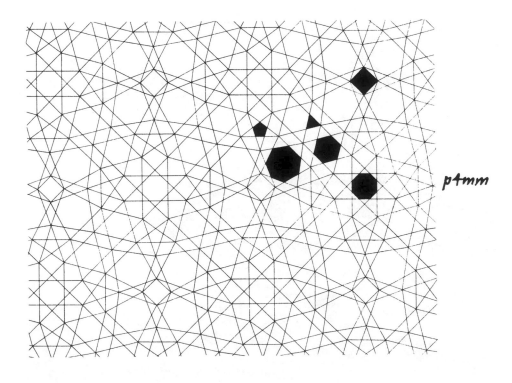

p4mm

Fig. 13.07

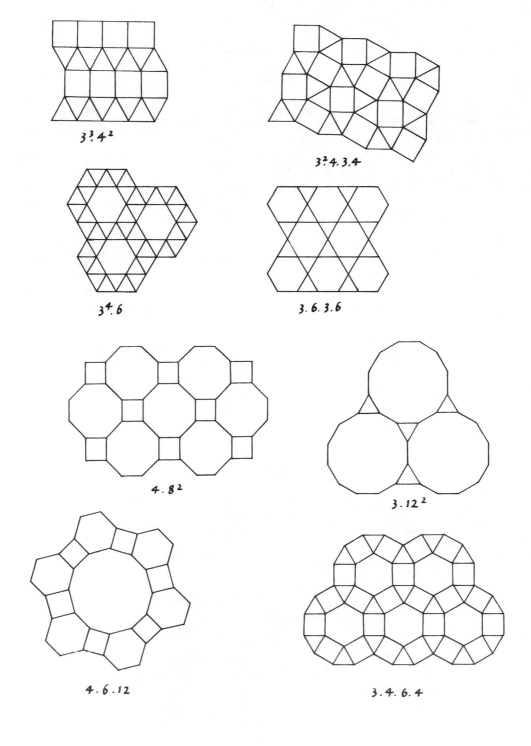

$3^3.4^2$

$3^2.4.3.4$

$3^4.6$

$3.6.3.6$

4.8^2

3.12^2

$4.6.12$

$3.4.6.4$

extensively by M.C. Escher in his plane-filling designs (Figs. 11.14 and 11.15, p. 66).

Tessellations

When a floor or wall is covered with tiles a repetitive and symmetrical pattern is often formed, straight edges being more common than curved ones. Such a division of the plane into polygons, regular or irregular, is called a *tessellation*. We have already had numerous examples, such as the cells and sub-cells of Fig. 12.04 (p. 73) and the grids of Figs. 13.05, 13.06. *Regular tessellations* are those consisting of regular polygons of one kind only. There are three of these, made up of equilateral triangles or squares or regular hexagons. In *semi-regular tessellations* polygons of more than one kind are used, but the pattern must be the same at each vertex. It can be proved that there are only eight of them, as shown in Fig. 13.07. The notation used here, known as *Schäfli's notation*, indicates the pattern at a vertex. Thus $3^2.4.3.4$ means that at each vertex there are two triangles, a square, another triangle and another square, in that order.

Removal of the restriction that the pattern must be the same at each vertex makes possible a substantial number of additional tessellations composed solely of regular polygons, as, for example, Fig. 13.08(*a*); and if, alternatively, it is admissible for the vertices of some polygons to lie on the sides of others, then further interesting tessellations such as those of Fig. 13.08(*b*) can be found. Variations in all these patterns can be produced by distorting the sides of the polygons in a systematic manner, as in Fig. 13.08(*c*). (See M. Kraitchik: *Mathematical Recreations*, pp. 199–207). It is surprising that only a few of these tessellations are in common use now, though many are found in Byzantine buildings.

To every regular or semi-regular tessellation there corresponds a reciprocal tessellation formed by the perpendicular bisectors of the sides of the polygons in the original. Thus the reciprocal of a triangular lattice is a hexagonal one

Fig. 13.08

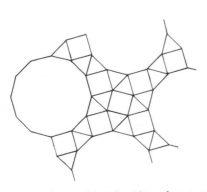

$12.4.3^2$ combined with $3^2.4.3.4$

(*a*)

6.3.1

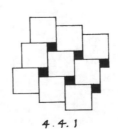

4.4.1

(*b*)

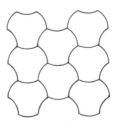

Distorted hexagons

(*c*)

(Fig. 13.09(a)) and the reciprocals of the tessellations $3^2.4.3.4$ (Fig. 13.09(b)) and $3^4.6$ (Fig. 13.09(c)) are patterns of congruent pentagons such as are often used for street paving in Moslem countries. A tessellation and its reciprocal are necessarily of the same symmetry type. Those of our examples are *p6mm*, *p4gm* and *p6* respectively.

Mosaic patterns having colour symmetry of any of the types in Table 12.1 (p. 79) can be derived from the corresponding diagrams of sub-cell arrangement. For symmetry of two colours (or two colour schemes) we must have a pair of sub-cells patterned in the same way but with different colouring. For simplicity we use two colours only, black and white, but each sub-cell may contain both, its 'pair' having the same colours reversed. A very simple pair, with reflected versions, is shown in Fig. 13.10. Using these with the sub-cell arrangements of Figs. 12.06, 12.07 (p. 74) we obtain the coloured cells shown in Fig. 13.11. These are the basic cells whose repetition produces the bicoloured symmetry patterns indicated by the symbols.

Patterns having multicoloured symmetry can be derived in the same way from the sub-cell arrangements of Figs. 12.09, 12.10 (pp. 75, 76). Using sub-cells of the same type as in Fig. 13.10, but with a sequence of four colours, we obtain the patterns of Fig. 13.12.

Fig. 13.09

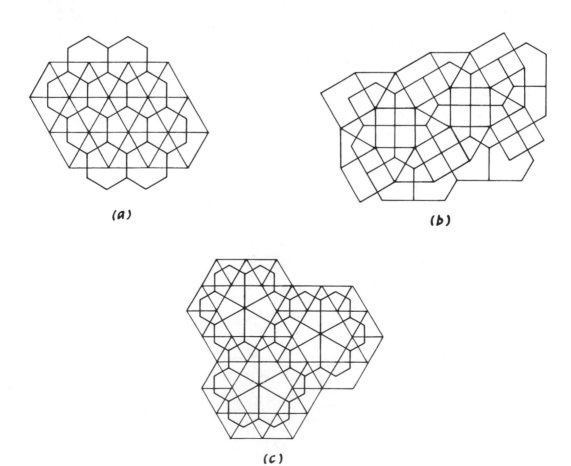

(a)

(b)

(c)

Fig. 13.10

Fig. 13.11

$p\,4\,g'm'$ $p\,4'g\,m'$ $p\,4'g'm$ $p'_n\,4g\,m$

Fig. 13.12

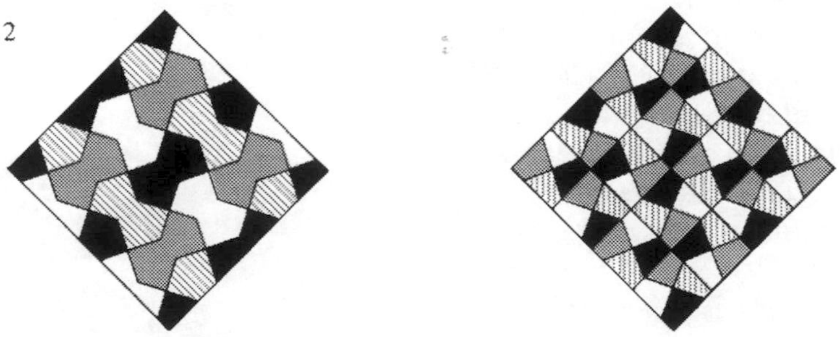

$p\,4^{(4)}mg^{(4)}$ $p\,4^{(4)}gg^{(4)}$

Automatic reproduction

There are various ways, some very ancient, for producing repetitive patterns automatically. The simplest of all is printing or imprinting. A die is cut with the form of the motif and this is systematically pressed into the clay or plaster being decorated. Between impressions the die may either be held in the same orientation or turned. If the surface of the die is flat it may be inked and used to produce a printed impression on paper or fabric, and often the pattern so produced is used as the basis for further work, such as carpet making, embroidery or other decorative crafts. It would even be reasonable to conjecture that symmetrical patterns first arose from the production process rather than from any desire on the part of primitive craftsmen for symmetry as such. In just the same way symmetry in nature may arise, not from the inclination of a creator, but rather as an inevitable consequence of the processes of growth, whether animate or in plant life or in crystals.

Cloth weaving and basket making must also produce diverse patterns, sometimes of very complex symmetry, solely as a result of the interleaving process necessary to obtain a strong fabric or artifact. As a consequence the work of primitive peoples often shows design symmetry of a high order and interest.

Consider, for example, the plaited straw pattern from the Sandwich Islands shown in Fig. 13.13, which has the coloured symmetry *c'1m*.

Fig. 13.13

c'11m

There is a method of producing symmetrical patterns which is used by children today and has been known to the Japanese and practised by them for many centuries. This is the cutting of folded paper. If a rectangular piece of paper is folded with a single crease down the middle and a piece cut from it, the part cut out and the remainder, when unfolded, both have reflexional symmetry about the line of the fold. When more than one fold is made, successive reflexions occur in the cut-out pattern. Two folds meeting at right angles produce a design of symmetry *2mm*, and if a third fold is made with the crease bisecting the angle between the other two the resulting pattern has symmetry *4mm* (Fig. 13.14). Similarly by arranging for the folds to meet in a point at angles of $180°/3$, $180°/5$, $180°/6$, etc., it is possible to produce patterns having symmetry *3m*, *5m*, *6mm*, etc.

If successive folds are parallel, the resulting pattern will have translational repetitions with parallel reflexion lines. A strip of paper, after folding in this manner, can be reduced to a rectangle, so producing the frieze pattern *r1m*. If it is first folded lengthwise the frieze is of form *r2mm*.

Similarly a sheet of paper can be folded by parallel creases into a strip and the strip folded further to make a rectangle, or various triangles, as in Fig. 13.15. From (*a*) is obtained the plane pattern *p2mm*, from (*b*) *n3m*, from (*c*) *p4mm*, and from (*d*) *p6mm*. If the shape cut from the folded paper has itself some suitable symmetry, further plane patterns can be obtained. Thus if in case (*a*) a piece having symmetry 2 is cut from the centre of the rectangle, the resulting pattern is of type *c2mm*; and if in case (*b*) the piece has symmetry 3, then *p31m* is obtained. Finally, if a sheet is folded to a square and a piece having symmetry 4 is cut from its centre, the pattern formed is of type *p4gm*. Thus the folded paper method can be used to produce seven of the plane symmetry types, all those in fact that have parallel reflexion lines in more than one direction.

Fig. 13.14

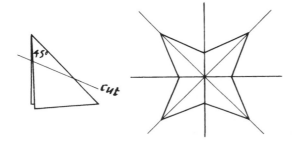

Fig. 13.15

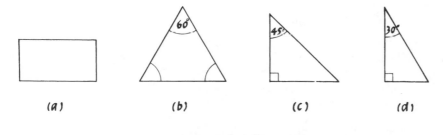

(*a*)   (*b*)   (*c*)   (*d*)

A more ephemeral method for producing plane repetitive patterns is to use mirrors, as in the kaleidoscope. The kind familiar as a child's toy has two mirrors only and can accordingly produce only a finite pattern, such as $3m$, $4mm$, $5m$ or $6mm$, according to whether the angle between the mirrors is 60°, 45°, 36° or 30°. A repetitive pattern can be formed if a third mirror is added, forming a triangle as in Fig. 13.15(b), (c) or (d). The pattern is then $p3m$, $p4mm$ or $p6mm$. With four mirrors forming a rectangle we can also obtain $p2mm$. If a pattern element having two-, three- or four-fold rotational symmetry is placed centrally, the kaleidoscope can give three more types, as with folded paper. But the kaleidoscope offers a further interesting possibility in that a third mirror can be added in a plane cutting across the line of intersection of the first two. A motif reflected in these three mirrors can form a pattern having three-dimensional point symmetry if the angles between the mirrors are suitable. In this manner, using one, two or three mirrors, it is possible to produce all the types of finite pattern that have intersecting reflexion planes, including such non-crystallographic types as $5m$ and $\bar{5}\,\bar{3}\frac{2}{m}$. An elegant demonstration with such a kaleidoscope is to place some mercury in the apex between the three mirrors: with appropriate angles between the mirrors, polyhedra are generated. If bent pieces of wire are reflected in the mirrors, stellated polyhedra can be formed.

It will be noted that for plane kaleidoscopes the triangle of mirrors, shaped as shown in Fig. 13.15(b), (c) or (d) is placed with its sides parallel to the reflexion lines of the pattern to be produced. A similar process can be applied in three dimensions, with the mirrors forming a tetrahedron whose faces are parallel with the reflexion planes of cubic space groups. In this way we can obtain patterns of types $P\frac{4}{m}\bar{3}\frac{2}{m}$, $F\bar{4}3m$ and $F\frac{4}{m}\bar{3}\frac{2}{m}$. (See Shubnikov and Koptsik, pp. 200−3.) With two parallel mirrors and three more perpendicular to them, forming a triangle as before, we obtain patterns corresponding to repetitions of the plane kaleidoscope patterns in a third dimension. These are $P\frac{2}{m}\frac{2}{m}\frac{2}{m}$, $P\frac{4}{m}\frac{2}{m}\frac{2}{m}$, $P\bar{6}m2$ and $P\frac{6}{m}\frac{2}{m}\frac{2}{m}$. Some forther groups coule be generated by placing in such a kaleidoscope a pattern element having rotational symmetry.

Computer-aided design

There is a third possible method for obtaining symmetrical designs which has not yet been exploited, though clearly it has a greater scope for development than any other. It has the capacity to produce any symmetrical pattern, in two or three dimensions, without restriction to certain particular types, as is the case with paper cutting and the kaleidoscope. A computer, when suitably programmed for a required symmetry group or groups, could be fed with an arbitrary element of pattern and would then automatically compute and display the reflexions, rotations and translations of the motif, as determined by the group in question.

For example, if the computer were programmed with the 17 wallpaper groups, the designer could feed in an element of pattern and, on demand, the machine would first display repetitions, to form a $p1$ pattern, on a net which could be distorted at will. At the touch of a switch, two-fold rotations could be introduced, to give $p2$. Adding reflexions, the designer could observe the effect of $p2mm$ and then of $p4mm$. In this way he could see displayed a wide range of patterns and pattern types from which he could select those suitable for his purpose.

At the time of writing, it is not known whether the programming of a computer in the manner indicated has yet been undertaken. A question of some interest would be whether it could be achieved with economy, in order to obtain rapidity of operation. With some ingenuity applied to the programming, a powerful hand-held programmable calculator could be operated in this way. The output could not of course be a visual display but only the coordinates of a set of points corresponding to a given point of the motif.

An effective method of writing the programmes could be based on the use of matrices, as outlined in Appendix 2.

PROBLEMS

Three-dimensional objects

1. State the symmetry types of
   (i)   the truncated tetrahedron, 3.6^2 ;
   (ii) the cuboctahedron, $(3.4)^2$;
   (iii) the snub cube, $3^4,4$;
   (iv) the icosidodecahedron, $(3.5)^2$.

2. Show how to colour the faces of a cube, using 1, 2 or 3 colours, with one colour only on each face, to give patterns of the following (uncoloured) types:
$\frac{4}{m}\overline{3}\frac{2}{m}$, $\frac{4}{m}\frac{2}{m}\frac{2}{m}$, $4mm$, $2mm$, $\frac{2}{m}\frac{2}{m}\frac{2}{m}$, $3m$, $1m$.
   Show also how to obtain patterns for the following bicoloured and tricoloured types:
$\overline{4}'m2'$, $\overline{3}'m$, $3^{(3)}$, $3^{(3)}\frac{2}{m}$.

3. To what series of symmetry types does each of the series of crystal forms in Fig. 13.16 belong?

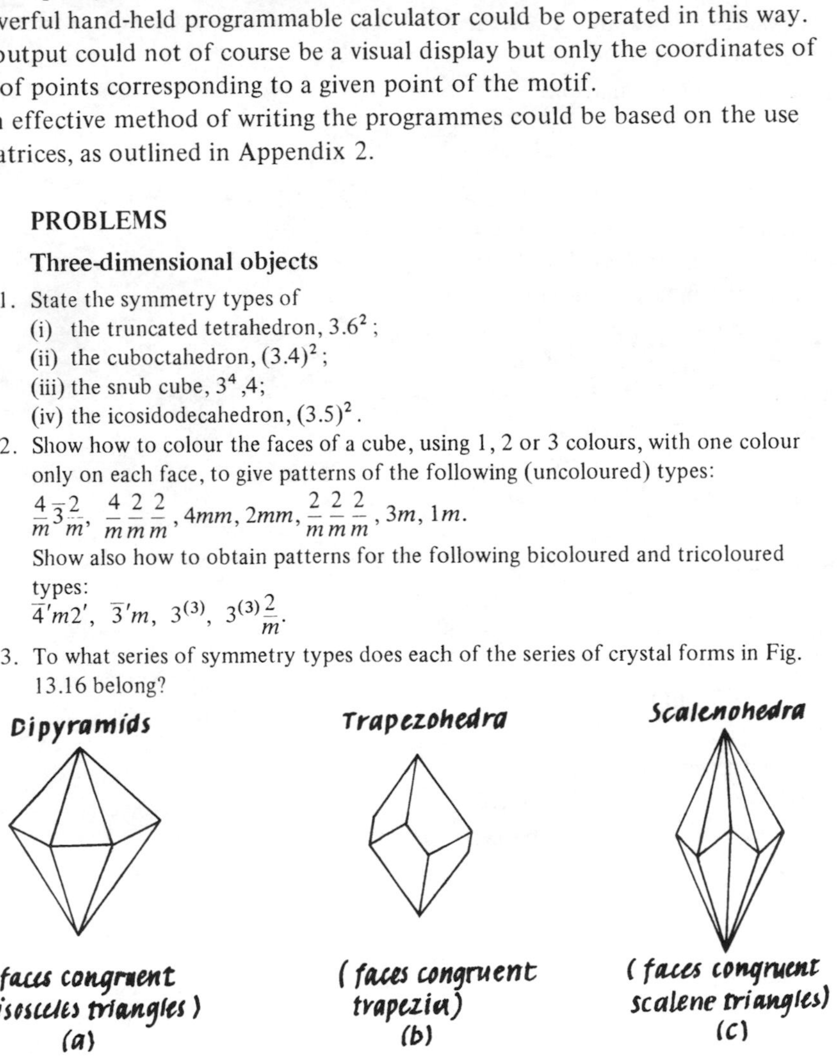

Fig. 13.16

Dipyramids

(faces congruent isosceles triangles)
(a)

Trapezohedra

(faces congruent trapezia)
(b)

Scalenohedra

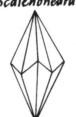

(faces congruent scalene triangles)
(c)

4. Classify the symmetry types of the following crystal forms:
   (i)   octahedron;
   (ii) pentagonal trisoctahedron (Fig. 13.17(a));
   (iii) pentagonal dodecahedron (Fig. 13.17(b));
   (iv) tetrahedron;
   (v)   pentagonal tristetrahedron (Fig. 13.17(c)).

Fig. 13.17

(24 pentagonal faces each having 3 sides of one length and 2 of another)
(a)

(12 pentagonal faces each having 4 sides of one length and 1 of another)
(b)

(12 pentagonal faces each having 3 sides of one length and 2 of another)
(c)

5. What are the symmetry types of the objects shown in Fig. 13.18?

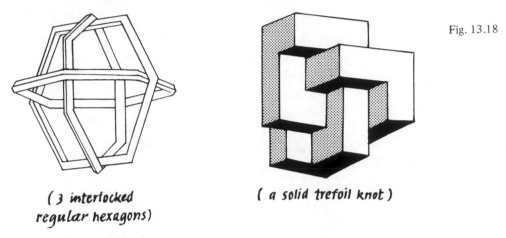

Fig. 13.18

(3 interlocked regular hexagons)

(a solid trefoil knot)

6. By colouring the appropriately sub-divided faces of the tetrahedron, octahedron and icosahedron, obtain patterns of the following bicoloured types: $\bar{4}'3m'$, $m'3m'$, $m'3m$, $m3m'$, $\bar{5}'\bar{3}m'$.

Tessellations

7. Identify the symmetry type of each of the tessellations of Fig. 13.07. How could the *p6mm* patterns be coloured to illustrate coloured symmetry types?
8. Use the pattern elements shown in Fig. 13.19 to make patterns of as many types as possible. Use each in turn alone, varying the orientation. Also use (*b*) with its reflected version.

(60° rhombuses)

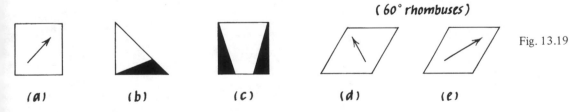

Fig. 13.19

(*a*)   (*b*)   (*c*)   (*d*)   (*e*)

9. From four squares it is possible to obtain the seven shapes shown in Fig. 13.20, including two enantiomorphic pairs. What types of pattern can be obtained from tessellations made (i) using any one of these shapes, (ii) using either pair of enantiomorphs?

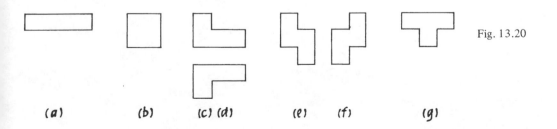

Fig. 13.20

(*a*)   (*b*)   (*c*) (*d*)   (*e*)   (*f*)   (*g*)

10. The shapes shown in Fig. 13.21 are each made from four equilateral triangles. What types of pattern can be obtained from tessellations made using any one of them?

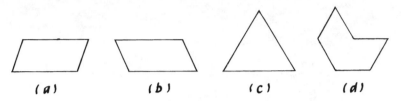

(a)   *(b)*   *(c)*   *(d)*

Fig. 13.21

11. A letter F can be formed from eight squares. Show how to make a tessellation from this shape, of symmetry $p2$. By colouring suitable elements obtain from it the tessellations $p'2$ and $p2'$.
12. Make tessellations using a Greek cross made of five squares or a Latin cross of six squares.
13. Investigate the colour symmetry mosaics obtainable from the patterns found from Figs. 13.20(g) and 13.21(d).
14. It is possible to fill space with rhombic dodecahedra (dihedral angles $60°$, $120°$). What is the symmetry type?

ANSWERS

1. (i) $\bar{4}3m$; (ii) $\frac{4}{m}\bar{3}\frac{2}{m}$; (iii) 432; (iv) $\bar{5}\,\bar{3}\frac{2}{m}$.

3. (a) $\frac{3}{m}2m \,(= \bar{6}2m), 4/mmm, \frac{5}{m}2m \ldots$; ($b$) $32, 432, 52, \ldots$; (c) $\bar{4}m2, \bar{6}m2, \ldots$.

4. (i) $\frac{4}{m}\bar{3}\frac{2}{m}$; (ii) 432; (iii) $\frac{2}{m}\bar{3}$; (iv) $\bar{4}3m$; (v) 23.

5. $\frac{2}{m}\bar{3}, 32$.

7. $c2mm, p4gm, p6, p6mm, p4mm, p6mm, p6mm, p6mm$;
   $p6'mm', p6'm'm, p6m'm', 6^{(3)}, 3^{(3)}$.

8. (a) $p1, p1m, p1g, c1m, p4, p2gg$.
   (b) $p2, p4; p1m, p1g, c1m, c2mm, p2mg, p2gg, p4mm, p4gm$;
   (c) $p1m, p2mm, p2mg, p4gm$.
   (d) $p2, c1m, p2mg, p3m, p6$;
   (e) $p2, c1m, p2mg, p3, p6$.

*9. (i) (a) $p1, p2mm, p2gg, p4mm, c2mm$;
        (b) $p4mm, p2, c2mm$;
        (c) $p1, p2, p4$;
        (e) $p2, p4$;
        (g) $p2, p2mg, c2mm, p4$.
    (ii) (c) and (d) $c2mm, p4gm$;
        (e) and (f) $p2mg, p2gg$.

*10. (a) $p2$. (a) and (b) $p2mg, p2gg$. (c) $p6mm, p2mg, p2$. (d) $p1m, p2mg, p6$.

12. $p4, p'4; p2, p2', p'2$ (see part 27 of Fig. 12.16, p. 000).

*13. For 9(g): $p'2, p2', p^{(n)}2; p2'g'm, p_a'2gm, p1g^{(n)}; c2'gm, p_n'2gm, p_n'2mm; p'4,$
     $p4', p4^{(4)}$.
     For 10(d): $p'1m, p^{(n)}1m; p2'm'g, p'2mg, p^{(n)}2mg; p6', p3^{(3)}, p6^{(3)}, p6^{(6)}$.

14. $P\frac{4}{m}\bar{3}\frac{2}{m}$.

*There are many answers to some of these questions. The selection given is by no means complete.

Part II The mathematical structure

A more mathematical approach must now be adopted in order to enumerate all the possible types of symmetry in each class. Basic to this procedure is the enunciation and proof of a number of theorems relating to symmetry movements and their interactions. In designing a symmetrical pattern we can choose certain basic movements, for example a rotation and two translations, and we apply them to a single unsymmetrical pattern element. Repetitions and combinations of these movements are equally applicable and thus the whole pattern is built up. The type of symmetry is characterized by the complete group of movements, the basic movements and all possible repetitions and combinations. When the pattern has been built up in this way we find that any one movement, applied to the pattern as a whole, leaves it unchanged. But we have to distinguish between the basic movements, which are at choice, and those that occur as a result of repetitions and combinations. That is why we begin with a study of movements and their interactions.

By this means we are able to show that, in two dimensions, the 7 types of discontinuous frieze pattern and the 17 types of discontinuous wallpaper pattern, as described in Part I, are the only ones. Continuing in three dimensions we show that there are 32, and only 32, types of discontinuous point symmetry, and on this basis we enumerate the types of three-dimensional repetitive pattern. We then consider groups containing continuous movements, and finally we discuss coloured symmetry, a development first studied during the present century.

14

Movements in the plane

Congruence and isometry

Two figures are congruent if to any point of one there corresponds a point of the other and the distance between any two points of the one is equal to that between the corresponding points of the other. It follows that corresponding angles are equal in magnitude; but the congruence may be direct or opposite, according as the corresponding angles are in the same sense or opposite senses. An isometry is a transformation or mapping such that any figure is mapped onto a congruent figure. As the congruence is direct or opposite, so is the isometry.

An isometry is determined by two pairs of corresponding points and a statement either that it is direct or that it is opposite.

Proof. Suppose that A', B' correspond to A, B respectively, where $A'B' = AB$ (Fig. 14.01). If C is any other point of the first figure, not in line with A and B, there are two and only two positions for C' such that $C'A' = CA$ and $C'B' = CB$. If the corresponding angles are in the same sense (as for C_1' in the figure) the isometry is said to be *direct*; if in opposite senses (as for C_2') it is *opposite*. Moreover when A', B' and C' are fixed there is only one possible new position for a fourth point D; for its distances from A', B' and C' are determined.

Movements

An isometry is the change from one position to another; it does not depend on the route by which the change is effected. A turn of 90° clockwise and one of 270° anticlockwise are, strictly speaking, different movements, but they have the same effect. They produce the same isometry, and for our purposes they are equivalent. We shall use the term *movement* to mean the change of position and we shall speak of *direct* and *opposite movements* according to the isometries they produce.

Fig. 14.01

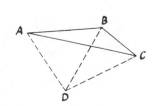
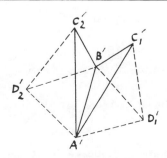

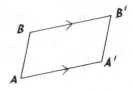

Fig. 14.02

Not every movement produces an isometry. For example, if every point is moved outwards from a fixed point to double its distance the result is a dilation. But here we are concerned only with those movements that are isometric and we shall use the word *movement* on that understanding.

Fig. 14.03

A *translation*, in which each point is moved a fixed distance in a fixed direction, is isometric; for if AA' and BB' (each representing the *translation vector*) are equal and parallel, then $A'B' = AB$ (Fig. 14.02). Moreover the isometry is direct, since the turns from the direction of AA' to those of AB and $A'B'$ are in the same sense.

A *rotation* is isometric; for if OA and OB (Fig. 14.03) are each turned through an angle a about O to the positions OA' and OB', it follows by congruent triangles that $A'B' = AB$. Moreover the isometry is again direct, since the turn θ from AB to $A'B'$ is not only equal to the turn from OA to OA' but in the same sense.

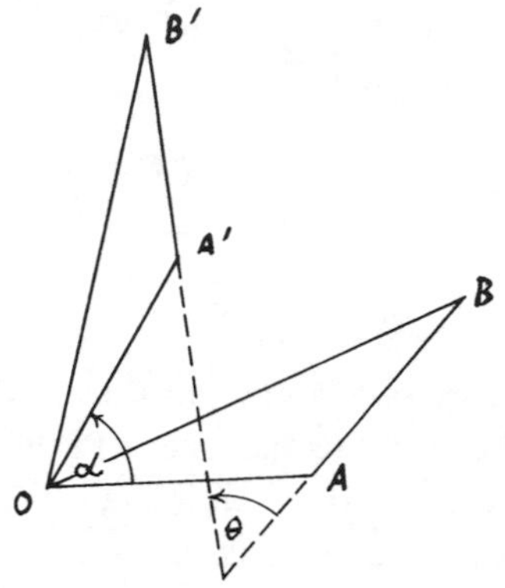

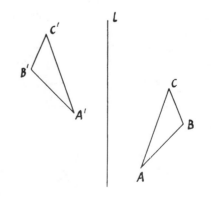

A *reflexion*, in which points $A, B, \ldots$ are replaced by their images $A', B', \ldots$ in a mirror line l (Fig. 14.04), produces an opposite isometry; for $A'B' = AB$ and the turn from $A'B'$ to l is equal and opposite to that from AB to l.

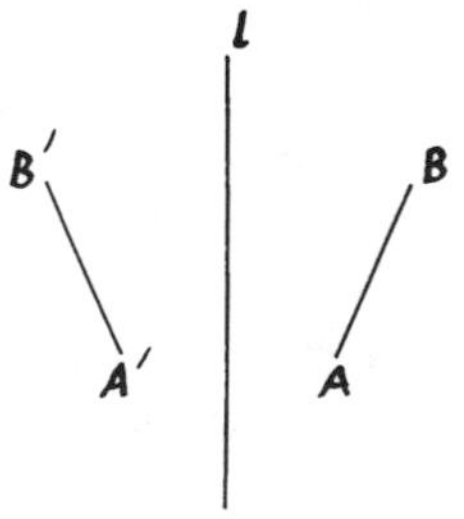

A *glide reflexion* is the combination of a reflexion in a line l and a translation parallel to l (Fig. 14.05). This too is an opposite isometry.

It will be shown that any isometry in the plane can be effected by one or other of these four movements. They are therefore the only movements in the plane that need be considered.

> *Direct movements*: A translation will be denoted by T, a rotation by S.
> *Opposite movements*: A reflexion will be denoted by M, a glide reflexion by G.

Fig. 14.04

When two movements are applied in succession the resultant will be denoted by a pair of letters, *the second letter referring to the first movement*. Thus TM means that a reflexion is followed by a translation. The resultant movement in such a case will be called the *combination* or *product* of the two movements. Thus with T parallel to the mirror line of M, the combination TM is a glide reflexion G and we may write $G = TM = MT$, this particular combination being commutative. A movement X repeated will be denoted by X^2 and the inverse of X by X^{-1}. Thus $M^2 = I$ (the identity) and $M^{-1} = M$.

Fig. 14.05

As an opposite movement changes the sense of all angles and a direct movement does not, it follows that the product of two direct movements, or of two opposite ones, is a direct movement; whereas that of a direct movement and an opposite one is an opposite movement. Thus two reflexions are together equivalent either to a rotation or a translation (as will be proved in detail below); but the product of a reflexion and a rotation is either another reflexion or a glide.

Isometries and movements

> *Any direct isometry can be effected either by a rotation or by a translation.*

Since two pairs of points determine a direct isometry, it is sufficient to prove that two points A, B can be brought to given new positions A', B', where $A'B' = AB$.

(i) Let the perpendicular bisectors of AA' and BB' intersect, if possible, at O (Fig. 14.06). Then $OA = OA'$ and $OB = OB'$ and triangles $OAB, OA'B'$ are congruent; hence angle $AOA' = $ angle BOB'. So a rotation about O would bring A, B to A', B'. But we have shown that a rotation produces a direct isometry and therefore the rotation about O produces the one determined by AB and $A'B'$.

(ii) If the perpendicular bisectors do not meet in a point, AA' and BB' are either parallel or in the same straight line. If they are in the same straight line, a translation would bring AB to $A'B'$. If they are parallel the four points form either a parallelogram or an isosceles trapezium (Fig. 14.07). If it is a parallelogram, AB could be brought to $A'B'$ by a translation; if an isosceles trapezium, a centre of rotation could be found by producing AB and $A'B'$ to meet at O.

> *Any opposite isometry can be effected either by a reflexion or a glide reflexion.*

It is again sufficient to deal with two points A, B and their new positions A', B'. Let l be a line drawn through the mid-point of BB' parallel to the bisector of the angle formed by AB and $A'B'$ (Fig. 14.08). Let A'' and B'' be the images of A and B in l. Then either A'' and B'' coincide with A' and B', in which case the movement is a reflexion, or $B'B''$ is parallel to l (by the intercept theorem), with $A''B''$ equal and parallel to $A'B'$, the movement being then a glide reflexion.

Angles of rotation

A translation does not alter the orientation of any line, i.e. in a translation a line AB moves parallel to itself (Fig. 14.02). In a rotation all lines are turned through the same angle (for, in Fig. 14.03, since angle $OA'B'$ = angle OAB, $\theta = a$.

When two rotations are combined the angles of rotation are additive, whether the rotations are about the same centre or not. For suppose that, in Fig. 14.09, O and Q are centres of rotation. There is one line OB_1 that turns about O through an angle a to reach the position OQB_2, and then about Q through an angle a' to reach QB_3, thus turning through $a + a'$ altogether. It follows by the above theorem that all lines of the figure turn through $a + a$.

Some combinations of movements
Two translations

Translations can be represented by vectors and are combined by vector addition. The combination (or product) of two translations is commutative.

Two reflexions

(i) If the mirror lines l and l', for reflexions M and M', are parallel and at distance d apart, the product $M'M$ (i.e. M followed by M') is a translation

Fig. 14.06

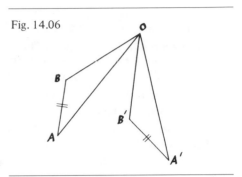

Fig. 14.08

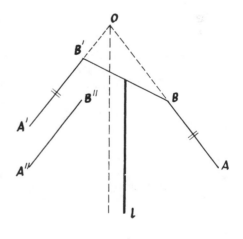

Fig. 14.07

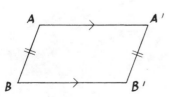

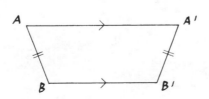

Fig. 14.09

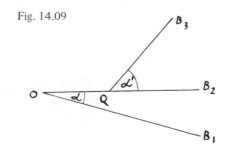

T of distance $2d$ in a direction perpendicular to them (Fig. 14.10). This combination is non-commutative, MM' being a translation of the same distance in the opposite direction. ($M'M = T, MM' = T^{-1}$.) The same diagram shows that $TM = M'$ and $M'T = M$ (or by algebra, $M'MM = TM$, therefore $M' = TM$, and similarly $M'M'M = M'T$); i.e. the product of a translation and a reflexion in a line perpendicular to it is a reflexion in a parallel line at a distance half that of the translation, on one side or the other, according to the order of operations.

(ii) If the mirror lines intersect at O, at an angle a, the product $M'M$ of the two reflexions is a rotation S about O of amount $2a$ (Fig. 14.11). The combination is again non-commutative ($M'M = S, MM' = S^{-1}$). From the same diagrams, or by algebra, $M'S = M$ and $SM = M'$; i.e. a rotation (through $2a$) followed by a reflexion (in l') is equivalent to a reflexion in a line (l) obtained by rotating l' half as much in the opposite sense; or, reversing the order, reflexion (in l) followed by the rotation gives reflexion in a line (l') obtained by rotation l half as much in the same sense.

Sylvester's Theorem*

An immediate consequence of the above is that if a triangle ABC has angles a, β, γ, as in Fig. 14.12, successive rotations, in the sense opposite to ABC, of amounts $2a$, 2β, 2γ, about A, B, C respectively, produce the identity: for the rotations are equivalent respectively to reflexions in AC and AB, in BA and BC, and in CB and CA, and these reflexions cancel one another out.

*As given by Sylvester (*Mathematical Papers* Vol. I, p. 160) this is a theorem in three dimensions, the rotation axes being concurrent, with ABC a spherical triangle. The two-dimensional version, which we use here, is a limiting case. The theorem is equivalent to Euler's Construction, to which similar remarks apply.

Fig. 14.10

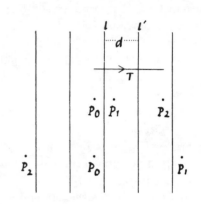

Fig. 14.12

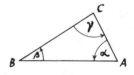

Fig. 14.11

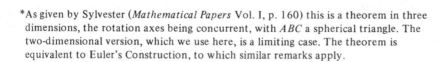

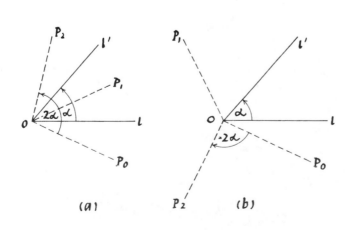

(a)   (b)

Fig. 14.13

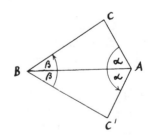

Fig. 14.14

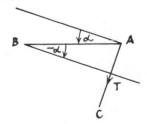

Fig. 14.15

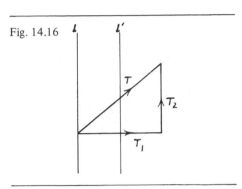

Two rotations (Euler's Construction)

It follows from the above that successive rotations about A and B of amounts $2a$ and 2β respectively are together equivalent to a rotation of amount -2γ about C, i.e. (in the plane) to one of $2a + 2\beta$ about C in the same sense as a and β. This important result may also be seen directly from Fig. 14.13: the rotation about A followed by that about B would bring C to C' and thence back to C, and as the product is a direct movement it can only be a rotation about C. If the order of the movements is reversed the product is a rotation about C'.

Fig. 14.16

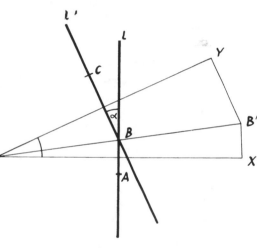

An important exception occurs when $2a + 2\beta = 0$, or a multiple of 2π. The resultant movement is then a translation, and it can be seen from Fig. 14.14 that, with β equal to $-a$, A moves to C and hence the translation is of distance $2AB \sin a$, in a direction inclined at $90°-a$ to AB. (This is for the rotation of $2a$ about A followed by that of $-2a$ about B. If the order is reversed the resultant translation is CA, not AC.)

Fig. 14.17

Rotation and translation

If S is a rotation of angle $2a$ about a centre O and OQ represents the translation T (Fig. 14.15), TS is an equal rotation about a centre C such that $CO = CQ$ and the turn from CO to CQ is $2a$: for the combined movement is a direct one and leaves C unchanged. If the order of movements is reversed, ST is an equal rotation about D, where $OQCD$ is a parallelogram.

Reflexion and translation

It was shown on p. 100 that if the translation T is perpendicular to the mirror line l of a reflexion M (Fig. 14.10) the resultant movement TM is a reflexion in a line l' parallel to l and distant from it half the translation distance. If T is not perpendicular to l (Fig. 14.16) it can be resolved into T_1 perpendicular to l and T_2 parallel to it. The resultant is then the combination of a reflexion in l' and the translation T_2, in short a glide reflexion along l'.

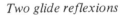

Fig. 14.18

Two glide reflexions

Let the glides G, G' be along the lines l, l' intersecting at B at angle a, and let the glide distances be AB and BC (Fig. 14.17). Let the perpendicular bisectors of AB, BC meet at O, and let the images of O in l and l' and in the point B be X, Y and B' respectively. The combined movement $G'G$ carries O to B' and thence back to O. But it is a direct movement and is therefore a rotation about O. Furthermore it carries A to B and thence to C, showing that the angle of rotation is $2a$.

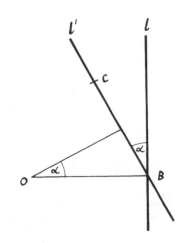

Reflexion and glide reflexion

This is a limiting case of the above. If M is a reflexion in l and G a glide reflexion of distance BC along l' (Fig. 14.18), the centre of the rotation GM is now the intersection of the perpendicular bisector of BC with a line through B perpendicular to l.

Rotation and reflexion

Let S be a rotation of angle $2a$ about O and M a reflexion in a line l. We know (p. 100) that if l passes through O the combination SM is a reflexion in another line through O. If not, we have the situation illustrated in Fig. 14.18, where

we know that $S = GM$ and hence $SM = GMM = G$. The combination is thus a glide reflexion along l'. (If the order of operations is reversed, l' is replaced by its mirror image in l. See p. 103.)

See p. 103.

Fig. 14.19

Rotation and glide reflexion

Let S be a rotation of angle $2a$ about O (Fig. 14.19) and let G be a glide reflexion along a line l. AB is the glide distance, marked off on l so that $OA = OB$. OB is rotated by S to the position OC. Then O is the centre of circle ABC and angle $BAC = a$. In the operation SG, AC is reflected to AC' and moved by translation to BC'', then rotated through $2a$, thus regaining its original direction. But as the rotation has carried B to C it will now be in a straight line with its original position. The combination SG is an opposite movement and it is therefore a glide reflexion of distance AC along AC. (Reversing the order of the movements, GS is a glide of the same distance along BC''. See p. 103.)

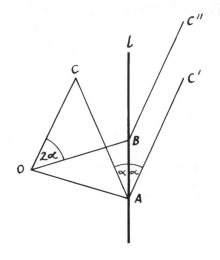

Transform of a movement

Movements may be related, by reflexions or rotations, in just the same way as the points of two figures. In Fig. 14.20, if O' is the reflexion of O in l, a clockwise rotation about O' could be regarded as the reflexion of an anticlockwise rotation about O. Again, in Fig. 14.21, if l' is the result of rotating l about O, a reflexion in l' can be regarded as the result of rotating a reflexion in l. If the reflexion in l is denoted by M and the rotation by S, the reflexion in l' is SMS^{-1}: for SMS^{-1} is an opposite movement and it leaves any point in l' unchanged. The line l' is described as $S(l)$ and the reflexion in l' could be $S(M)$, but it is usually written SMS^{-1}.

Fig. 14.20

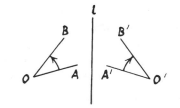

Similarly, in Fig. 14.20 if the rotation about O is S and the reflexion is M, O' is described as $M(O)$ and the rotation about O' is MSM^{-1}: for MSM^{-1} is a direct movement leaving O' unchanged, and therefore it is a rotation about O'. Moreover, if S changes OA to OB and M changes the triangle OAB to the oppositely congruent triangle $O'A'B'$, MSM^{-1} changes $O'A'$ to $O'B'$, a rotation equal and opposite to that from OA to OB.

Fig. 14.21

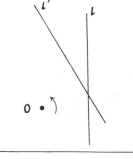

These are particular cases of a general relationship:

> If X and Y are any transformations, YXY^{-1} is a new transformation, the result of applying Y to X. It is called the transform of X by Y.

Suppose that X changes $P_0 \ldots P_n$ to $Q_0 \ldots Q_n$ (Fig. 14.22), and that Y changes $P_0 \ldots P_n$ to $P_0' \ldots P_n'$ and $Q_0 \ldots Q_n$ to $Q_0' \ldots Q_n'$. Then YXY^{-1} moves P_0' to P_0 and thence to Q_0 and finally to Q_0'.

It remains to interpret YXY^{-1}. If Y and X are both isometric, so is YXY^{-1},

Fig. 14.22

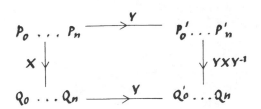

the isometry being direct if X is direct and opposite if X is opposite (quite independently of Y). If X is a rotation, leaving one point O unchanged, YXY^{-1} leaves the corresponding point O', i.e. $Y(O)$, unchanged. So a rotation about O becomes a rotation about $Y(O)$. Moreover it is a rotation of the same amount, since a triangle OP_1Q_1 is changed by Y to the congruent triangle $O'P_1'Q_1'$; but it will be in the opposite sense if the transformation Y is indirect.

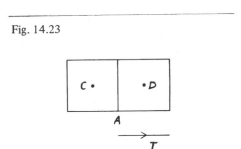

Fig. 14.23

If X is a reflexion, leaving all points of a line l unchanged, YXY^{-1} leaves all points of $Y(l)$ unchanged; so YXY^{-1} is a reflexion in $Y(l)$. Again, if X is a translation, so that all lines P_rQ_r are equal and parallel, then the corresponding lines $P_r'Q_r'$ are all equal and parallel to one another, and equal to P_rQ_r (though not necessarily parallel to it). Finally, if X is a glide reflexion, i.e. a combination of reflexion in a line l and a translation parallel to l, YXY^{-1} is the corresponding combination of a reflexion in $Y(l)$ and a translation of the same amount parallel to it.

The conclusion is that the transformation YXY^{-1} is always of the same kind as X and may be interpreted as the result of applying Y to X.

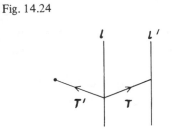

Fig. 14.24

Order of operations

Among its many applications, the above theorem provides a convenient way of ascertaining the effect of reversing the order of two operations. If $XY = Z$, $YX = YXYY^{-1} = YZY^{-1}$. So, if we know the effect of XY, we have only to apply the transformation Y to obtain the effect of YX. For example, if C and D are centres of the squares in Fig. 14.23, with T the translation shown and Q_A, Q_C, Q_D quarter-turns (anticlockwise) about A, C, D respectively, and if we know that $Q_A T = Q_C$, we can deduce that $TQ_A = TQ_CT^{-1} = Q_D$. Or again, if we know that $TQ_A = Q_D$, we can deduce that $Q_A T = Q_A Q_D Q_A^{-1} = Q_C$.

It is evident that movements are not usually commutative. $YX = XY$ only if $YXY^{-1} = X$. This happens, for example, when X and Y are both translations, or when X is a translation and Y is a reflexion in a line parallel to it, or when X is a reflexion and Y is a half-turn about a point on the mirror line.

Interaction of movements

The importance of the theorem on p. 102 is that if transformations X and Y both leave a pattern unchanged, so do the transformations XYX^{-1} and YXY^{-1}. Examples belong properly to the chapters that follow, but some of particular note are given here in anticipation.

    (i) *Translation and rotation.* If a pattern allows a translation T and a rotation through $2\pi/n$ about a point O, it must also allow similar rotations about the points $T^r(O)$ and translations inclined to T at angles $2\pi/n$, where r is any integer. Thus in a wallpaper, if there is rotational symmetry about one point of the net, so is there about any of the other points; and, as the translation can be rotated, the net itself must allow the rotation about any of its points.

    (ii) *Translation and reflexion.* If there is a translation T and a reflexion M in a line l (Fig. 14.24) there must also be reflexions in parallel lines $T^r(l)$ (e.g. l' in the figure) and a translation MTM^{-1}, i.e. T' in the figure. In a wallpaper, for example, a reflexion line is repeated along with the rest of the pattern. Moreover the translations must include T' and hence the net must allow reflexions in lines parallel to l through any of its points.

Two special cases are of note: if T is parallel to l, $T' = T$, and if T is perpendicular to l, $T' = T^{-1}$; so in either case the reflexion produces no new translation.

(iii) *Rotation and reflexion.* If there is a rotation S through $2\pi/n$ about a point O and a reflexion M in a line l through O, there must also be reflexion lines through O inclined to l at angles $2\pi r/n$, where r is any integer.

Intermediate reflexion lines

The set of reflexion lines in (iii) above are related by the rotation. We know from the theorem (ii), p. 100, that there will be another set of reflexion lines bisecting the angles formed by the first set (Fig. 14.25). These lines too are related to one another by the rotation.

Similarly in (ii) there is a glide reflexion line midway between l and l' (Fig. 14.26) and in fact a succession of such lines, related to one another by the translation T. In the special case in which T is perpendicular to l the intermediate lines are lines of simple reflexion.

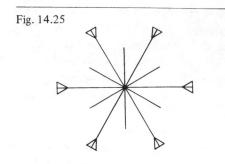

Fig. 14.25

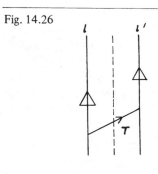

Fig. 14.26

15

Symmetry groups. Point groups

A *group of movements* consists of a set of movements satisfying the following conditions:

1. The product of any two movements in the set, or of any movement with itself, is a movement in the set.
2. The identity is included as a movement.
3. For every movement there is an inverse, a member of the set such that the product of the movement and its inverse (in either order) is the identity.
4. For three successive movements the associative law applies.

For example, if S is a rotation of amount $2\pi/3$, S^2 is one of $4\pi/3$ about the same centre and S^3 is a complete turn of 2π, equivalent to no movement at all. We say that $S^3 = I$, the identity. Two movements are counted as equivalent, for our purposes, if they bring a pattern to the same final position. Thus S^4 is equivalent to S, S^5 to S^2, and so on. Then S, S^2 and I form a group: for S and S^2 are inverses and I is its own inverse, and all possible products are shown in Table 15.1. It is easy to verify too that the associative law is always satisfied by

Table 15.1

		1st movement		
		I	S	S^2
2nd movement	I	I	S	S^2
	S	S	S^2	I
	S^2	S^2	I	S

these movements. In this particular group every product is commutative and the group is said to be *commutative* or *Abelian* (after the Norwegian mathematician Abel, 1802–1829).

The group may also be described as *discontinuous*, in the sense that it consists, apart from the identity, of movements that cause finite changes of position. It is with such movements that we are mainly concerned in this book, though patterns allowing continuous change have been considered in Chapter 9 and will appear again in Chapter 26.

For any finite group of movements it is possible to pick out certain basic movements of which all the rest are repetitions or combinations. A group having a single basic element, as, for example, the one just considered, is called a *cyclic* group. A group based on a reflexion, with or without a rotation about a point on the mirror line, is called *dihedral*. The simplest of these, consisting only of a reflexion M and the identity I, with a group table as shown in Table 15.2, is both dihedral and cyclic. If, however, there is a rotation as well as a reflexion, there are further movements obtained by combinations. For example, if Q is a quarter-turn about a point O, and M is a reflexion in a line through O, the group also contains rotations Q^2, Q^3 and Q^4 $(= I)$, and reflexions QM, Q^2M and Q^3M in lines through O at angular intervals of $45°$. These eight movements combine with one another as shown in Table 15.3.

Table 15.2

	I	M
I	I	M
M	M	I

Table 15.3

	I	Q	Q^2	Q^3	M	QM	Q^2M	Q^3M
I	I	Q	Q^2	Q^3	M	QM	Q^2M	Q^3M
Q	Q	Q^2	Q^3	I	QM	Q^2M	Q^3M	M
Q^2	Q^2	Q^3	I	Q	Q^2M	Q^3M	M	QM
Q^3	Q^3	I	Q	Q^2	Q^3M	M	QM	Q^2M
M	M	Q^3M	Q^2M	QM	I	Q^3	Q^2	Q
QM	QM	M	Q^3M	Q^2M	Q	I	Q^3	Q^2
Q^2M	Q^2M	QM	M	Q^3M	Q^2	Q	I	Q^3
Q^3M	Q^3M	Q^2M	QM	M	Q^3	Q^2	Q	I

Symmetry groups

A symmetrical figure is one that remains invariant under certain movements. For any particular figure the complete set of all such movements, if the identity is included, necessarily forms a group: for (i) the combination of any two such movements leaves the figure invariant and is therefore one of the set, (ii) each movement has its inverse in the set, and (iii) the associative law applies, because a movement of a point P_0 to P_1, to P_2 and to P_3 in succession can be regarded equally well as the movement from P_0 to P_1 followed by P_1 to P_3, or as P_0 to P_2 followed by P_2 to P_3.

Point groups

A *point group* is a group of movements all of which leave one point unchanged. The group shown in Table 15.3 is an example. Of the four possible movements in the plane, translations and glide reflexions move every point. A point group in two dimensions can therefore consist only of rotations about the point, reflexions in lines through the point, and the identity.

We consider first the groups containing only rotations (with of course the identity). For a discontinuous group there must be a smallest angle of rotation. Let this be S, a rotation through an angle a. Then $a = 2\pi/n$, where n is an integer: for if not, $2\pi/a$ will lie between two integers $(n - 1)$ and n. Then $n - 2\pi/a < 1$, and $na - 2\pi < a$. Hence the movement S^n is equivalent to a turn of less than a, contradicting the hypothesis.

The group therefore consists of $S, S^2, S^3, \ldots, S^{n-1}$ and S^n $(= I$, the identity). This is a cyclic group and clearly there are an infinite number of such groups, some of which are illustrated in Fig. 15.01. The first of these groups, with $n = 1$, consists of the identity only. It represents no movement at all. We can add to it a single reflexion line, as illustrated in Fig. 15.02; but the theorem on p. 99 shows that we cannot have more than one reflexion line without also having either translations or rotations. So further point groups must include rotations and we need only consider the addition of reflexions to the cyclic groups 2, 3, 4, ... Moreover, by the same theorem, if we add to the cyclic group n a reflexion in a line through the centre of rotation, we must also have reflexions in a set of lines at angular intervals π/n. This means n reflexion lines in all, as shown in Fig. 15.03 for $n = 1, 2, 3, 4, 5, 6$. When n is even these lines divide into two sets, the first set consisting of the original reflexion line and others related to it by the rotation, and the second set consisting of the intermediate reflexion lines bisecting the angles formed by the first set. When n is odd, the two sets coincide. This explains why the symbols have only one m when n is odd, but two, in the full form, when n is even. The shortened forms, shown in parentheses, omit what follows automatically from the theorem.

The symmetry group of a finite pattern is necessarily a point group. This will be proved in Appendix 1, but it may be noted here that a group containing a translation or a glide reflexion would lead, by repetition of the movement, to an infinite pattern, either a frieze or a wallpaper. Indeed a frieze may be regarded as built up from a finite pattern by applying a translation or a glide; and a wallpaper by applying two such movements. But the symmetries of the finite pattern will not necessarily be symmetries of the frieze or wallpaper. Thus a Tudor rose has pentad rotation, but a wallpaper in which it occurs cannot have that symmetry. The symmetry group of the frieze or wallpaper consists of those movements that leave the whole pattern invariant. It will be shown in the following chapters that the only point groups that can be part of a frieze group are the cyclic groups 1 and 2, and the dihedral groups $1m$ and $2mm$; and that the only ones that can be part of a wallpaper group are the cyclic groups 1, 2, 3, 4 and 6, and the dihedral groups $1m$, $2mm$, $3m$, $4mm$ and $6mm$.

Fig. 15.02

Fig. 15.01

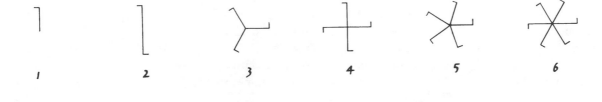

Fig. 15.03

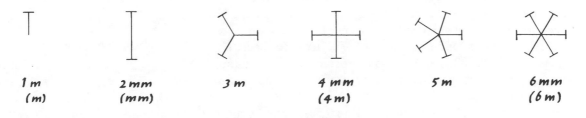

16

Line groups in two dimensions

A *line group* is a group of movements that includes translations in one and only one direction. We call that direction *longitudinal* and the perpendicular direction *transverse*. The simplest line group consists solely of translations T^r, where r is an integer and T is the shortest translation. (For the present we ignore the symmetry group of a uniform continuum, which alone has no shortest translation.)

A *frieze pattern*, described in Part I as built up from a motif or finite pattern by applying a translation or a glide, may now be redefined as a pattern in two dimensions invariant under the movements of a line group. Such a pattern is necessarily infinite. If any point O of the pattern is taken as origin, the points $T^r(O)$ form a *row* (Fig. 16.01). The pattern may be divided into *cells* by parallel lines spaced at the same intervals. (Usually it is convenient to place them midway between the points of the row.) Each cell contains an *element of the pattern* (a *motif*), the repetition of which makes up the whole frieze.

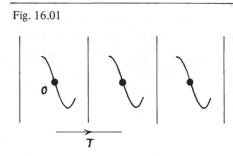

Fig. 16.01

The group may contain other movements, namely rotations, reflexions and glide reflexions. These, however, are limited by their interactions with the translation T.

(i) *Rotations.* A rotation s implies a further translation sTs^{-1}, and this must be T^r for some value of r. In fact it must be either T or T^{-1}, since it is a translation of the same distance as T. Therefore s is either a whole turn (equivalent to the identity) or a half-turn. A pattern to which the half-turn is applicable is shown in Fig. 16.02. A half-turn about O and a translation T, of distance d, combine to produce a half-turn about a centre distant $\frac{1}{2}d$ from O. There are thus two sets of half-turn centres, each set being spaced at intervals d. If the boundary lines of the cells are taken through either of these sets (Fig. 16.02(a), (b)) the element of pattern in the cell has the half-turn symmetry.

(ii) *Reflexions.* The possibilities are again limited by the interactions with the translation: for a reflexion m implies a translation mTm^{-1}, of the same distance as T, and this must be T or T^{-1}. Hence the reflexion lines can only be longitudinal (Fig. 16.03(a)), or transverse (b), or both (c). There can be only one reflexion line in the longitudinal direction, for two would imply a transverse translation; and it must pass through any half-turn centres, otherwise there would be a parallel reflexion line and hence a transverse translation. It is desirable that the division into cells should be so arranged that each cell may exhibit, if possible, the rotations and reflexions of the pattern. To ensure this the boundary lines of the cells in (b) and (c) should coincide either with the original reflexion lines or the intermediate set. In (c) the two reflexions combine to give half-turn rotations.

(iii) *Glide reflexions.* If g is a glide reflexion, g^2 is a translation of double the distance and in the same direction. It must be one of the translations T^r. If r is even (= $2q$, say), the glide is a combination of a reflexion with a translation such as can occur in Fig. 16.03(a) or (c); but if r is an odd number ($2q + 1$), the glide distance is ($q + \frac{1}{2}$) units (taking the translation distance of T as one unit). The group thus includes a basic glide reflexion of distance half a unit in the direction of the translation (Fig. 16.04). It may also include reflexions in transverse mirror lines, as in Fig. 16.04(b), in which case there are centres of half-turn rotation at distances of $\frac{1}{4}$ unit from the reflexion lines. The cell boundaries may be taken to coincide with either set of reflexion lines, or they may be drawn through either set of rotation centres.

The limitations on line groups may be summarized by saying that the symmetry movements must be derived from those of a row, $r2mm$.* That is to say, there may be half-turns about centres spaced at half-unit intervals along the line of the row, or reflexions in transverse mirror lines, again at half-unit intervals, or a reflexion in the line of the row itself as a longitudinal mirror line; or there may be all three of these symmetries (for any two imply the third). There may also be glides, these being combinations of a longitudinal reflexion with certain translations, namely whole units of the basic translation, as in Fig. 16.03(a) or (c), or odd multiples of half a unit, as in Fig. 16.04.

Thus from $r2mm$ we may derive seven line groups, five by selecting subgroups from $2mm$ and two by replacing the longitudinal reflexion by a glide. The seven are shown together in Fig. 16.05.

*As before, we use r to indicate translation as represented by the row, and 2 for the half-turns, with the first m for transverse and the second for longitudinal reflexion.

Fig. 16.02

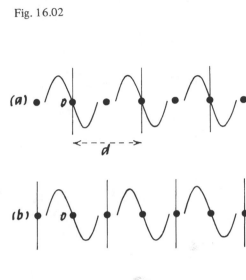

(a)

(b)

Fig. 16.03

(a) (b) (c)

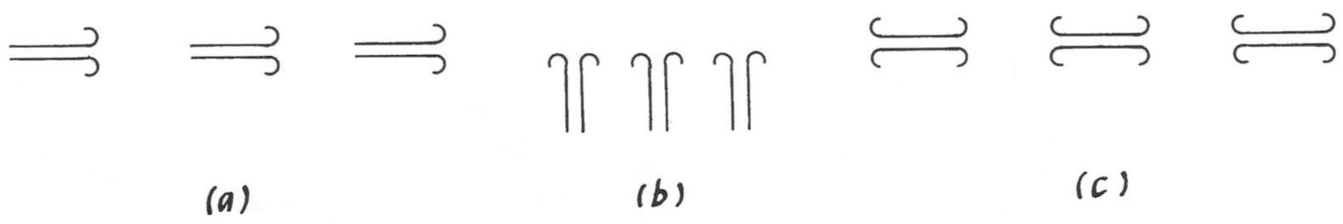

Fig. 16.04

|←— 1 unit —→|

(a) (b)

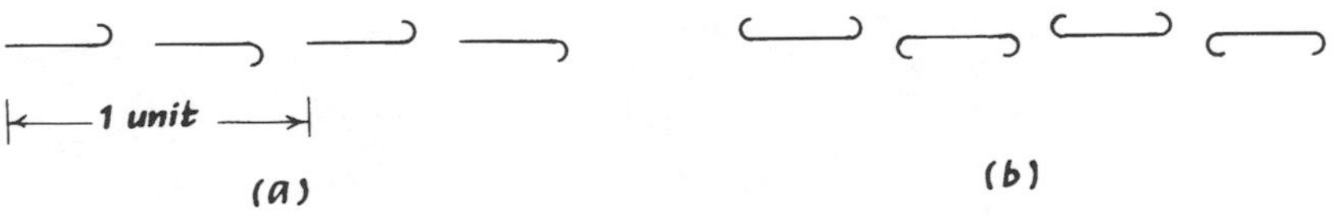

Fig. 16.05

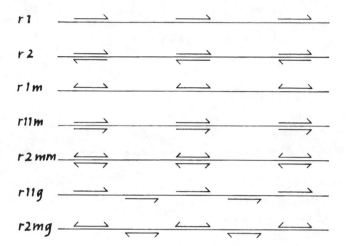

17

Nets

A *net* is an array of points determined by an origin O and two translation vectors T_1, T_2, not in the same direction. It consists of all points reached from O by the translations $T_1^p T_2^q$, where p and q are integers. Such translations form a group. The point reached from O by the translation $T_1^p T_2^q$ can be described by the coordinates (p, q). If the same translation is applied to any other point (p', q') of the net, it is taken to $(p' + p, q' + q)$, another point of the net. The net is therefore left unchanged by the translations of the group.

The points $(0, 0)$, $(0, 1)$, $(1, 0)$, $(1, 1)$ form a parallelogram which can be regarded as a unit cell, the whole net being mapped out by contiguous cells of that shape and size. Alternatively the cells may be parallelograms of the same shape and size but centred at points of the net. Every net has diad rotation symmetry about the centres of the parallelograms, the vertices and the midpoints of the sides (Fig. 17.01).

Some nets have further symmetries and there are in fact five types in all, distinguished by the symmetries they possess. To consider the possibilities we note first that any net can be built up from a row by applying a second translation vector. This produces a series of parallel rows identical with the original except for position. We have to consider what symmetries can be produced by different ways of choosing the second vector.

The symmetries to be considered are rotation, reflexion and glide reflexion. Rotations can only be diad, triad, tetrad or hexad. For suppose C_1 and C_2 are two similar centres of rotation as near together as possible (Fig. 17.02(a)). Then a rotation about C_1 of less than 60° (i.e. $2\pi/n$, where $n > 6$) would imply another centre C_3 nearer to C_2 than the supposed minimum distance. More precisely, if S_1 and S_2 are the rotations about C_1 and C_2 respectively, $S_1 S_2 S_1^{-1}$ is a similar rotation about the point $S_1(C_2)$, i.e. about C_3. Similarly if n were equal to 5, turns of $\pm 2\pi/5$ about C_1 and C_2 (Fig. 17.02(b)) would produce centres C_3 and C_4 nearer together than C_1 and C_2.

Every net has diad rotation, so we need only consider triad, tetrad and hexad.

Fig. 17.01

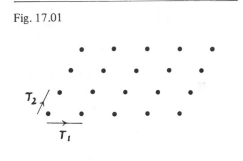

Fig. 17.02

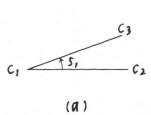
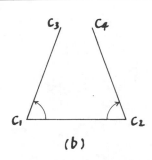

(a)   (b)

Triad rotation

Suppose that triad rotation about C brings a point A_1 of the net to A_2 and brings A_2 to A_3. C itself may or may not be a point of the net (Fig. 17.03(a), (b)). The net is defined either by the point C and the vectors CA_1, CA_2, or (if C is not a point of the net) by the point A_1 and the vectors A_1A_2, A_1A_3. In either case the net consists of equilateral triangles (Fig. 17.04). It has triad rotation symmetry about the vertices and centres of the triangles as well as diad rotation about the vertices and the mid-points of the sides; and this implies hexad rotation about the vertices.

For this net the unit cell could be the usual parallelogram, but alternatively a hexagonal cell may be used, as in Fig. 17.05. This has the advantage that the centres of triad and hexad rotation are more clearly shown.

Hexad rotation

For hexad symmetry the centres of rotation must be points of the net; for if rotation about C brings A_1 to A_2 and A_2 to A_3 (Fig. 17.06), the translation A_2A_3 would bring A_1 to C. The net is as before.

Tetrad rotation

For tetrad symmetry a centre C may or may not be a point of the net (Fig. 17.07(a), (b)), but in either case the net consists of squares, the side of a square being CA_1 or A_1A_2.

Reflexion

If l is a mirror line, let A_1 be a point of the net as near as possible to l, but not on it. Then if A_1 reflects to A_2, the two points are points of a row in a line perpendicular to l. The point of intersection with l may or may not be a point of the row (Fig. 17.08(a), (b)), but in either case the row has reflexion in lines parallel to l through any of its points and also through points half-way between them (Fig. 17.09). For the whole net to have reflexion about these lines the parallel rows must be arranged in one or other of the ways shown in Fig. 17.10. Thus the unit cell is either a rectangle or a rhombus; but it is convenient to regard the net of Fig. 17.10(b) as built up from centred rectangles rather than from rhombuses.

Both these nets have reflexion lines in two perpendicular directions, and this was to be expected, since reflexion in a line and a half-turn about a point of the line combine to produce a second reflexion in the perpendicular direction. Reflexions in other directions need not be considered, because reflexions in intersecting lines always imply a rotation, and all possible rotations have already been discussed.

Fig. 17.03

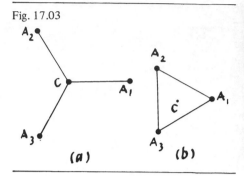

Fig. 17.04

Fig. 17.05

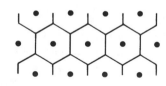

Fig. 17.06

Fig. 17.07   Fig. 17.08   Fig. 17.09

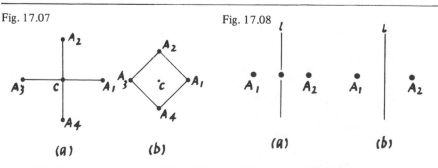

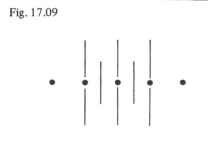

Glide reflexions

Suppose that a glide reflexion in a line l moves A_1 to A_2 and A_2 to A_3 (Fig. 17.11). If the pattern is continued by applying the translations $A_1 A_2$ and $A_2 A_3$ a centred rectangular net is obtained. Thus it appears that if a net has glide reflexion in any line it also has simple reflexion in parallel lines.

There are therefore only five types of net, as shown in Fig. 17.12. The fourth one is called 'hexagonal' because it has hexad rotational symmetry, as well as triad and diad. Reflexion lines of these nets are shown in Fig. 17.13, with broken lines for glide reflexions. In the symbols the prefixes p and c indicate primitive and centred nets respectively, and the letters at the lower level indicate reflexions or glides in intermediate lines.

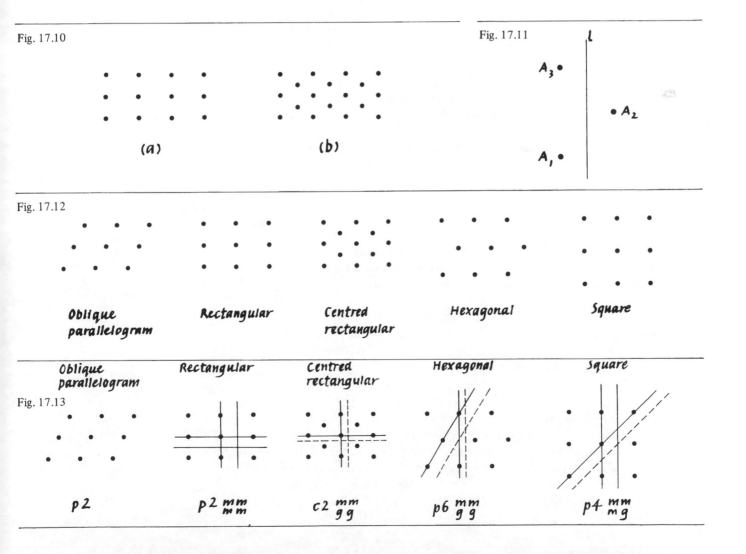

Fig. 17.10

(a) (b)

Fig. 17.11

Fig. 17.12

Oblique parallelogram   Rectangular   Centred rectangular   Hexagonal   Square

Fig. 17.13

Oblique parallelogram   Rectangular   Centred rectangular   Hexagonal   Square

$p2$ $p2\frac{mm}{mm}$ $c2\frac{mm}{gg}$ $p6\frac{mm}{gg}$ $p4\frac{mm}{mg}$

18

Plane groups in two dimensions

It was stated in Part I that there are just 17 types of 'wallpaper' pattern, i.e. of discontinuous plane pattern having repetitions in more than one direction. In this chapter we seek to prove that assertion. A pattern is characterized by its symmetry group, the group of movements that leave the pattern invariant. We have to prove that such groups fall into 17 distinct types.

Fig. 18.01

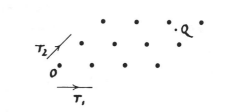

By 'distinct types' we mean types representing different sets of symmetry properties. Group structure is not by itself a sufficient criterion. For example, the rotation group 2, consisting of the identity and a half-turn, has the same structure as the group m, containing the identity and a single reflexion. They are isomorphic, but we regard them as distinct.* For symmetry groups to be of the same type not only must they be isomorphic but corresponding members must be movements of the same kind, a rotation corresponding to a rotation, a reflexion to a reflexion, and so on.

A discontinuous plane group is one containing translations in more than one direction in a plane, none of the movements being continuous. Let T_1 and T_2 be translations in the group, not in the same direction and of shortest possible distances. Then $T_1{}^p T_2{}^q$, where p and q are integers, are translations forming a sub-group, and we can prove that this sub-group includes all translations of the group.

Proof. Applied to any point O, the translations of the sub-group determine a net (Fig. 18.01). Suppose that T is any translation in the group, and that it carries O to a point Q. Then either Q is a point of the net, in which case T is a member of the sub-group; or its distance from the nearest point of the net is less than the translation distance of T_1 or T_2, and there is then a translation in the group shorter than T_1 or T_2, contrary to hypothesis. So T is a member of the sub-group.

As well as the translations the group may contain rotations about various centres and reflexions and glide reflexions in various lines (these lines not necessarily passing through the rotation centres). We proceed to show how these movements can be related to a point group and a net. The five possible nets and the ten possible point groups will then be limiting factors in the choice of plane groups.

Given any plane group, we first map the various rotations s into corresponding rotations S about an arbitrary point P. We then map the reflexions m and glide reflexions g into reflexions M in lines through P parallel to the original reflexion or glide lines. (This is illustrated in Fig. 18.02, where the rotations are half-turns and there are both reflexions and glide reflexions in two perpendicular directions.) The movements S and M, with the identity, then form a group.

*For further examples see Index of Groups.

Fig. 18.02

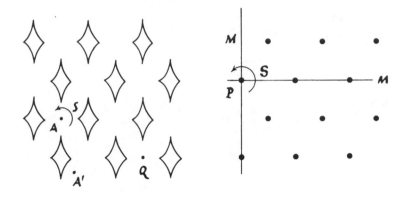

Proof. Rotations in the plane group through angles a, β, whether about the same centre or not, combine to give one of angle $a + \beta$, and this also happens with the corresponding rotations about P. Again, two reflexions or glide reflexions, or one of each, if they are in parallel lines, combine to give a translation, and the corresponding movements give the identity; but if they are in lines intersecting at angle a they combine to give a rotation through $2a$, and the corresponding reflexion lines through P do the same. For other combinations similar remarks apply. It is also easily seen that every movement has its inverse and that the associative law applies, and therefore the movements at P form a group.

We now construct a net, with origin at P, by applying the translations T_1, T_2 of the plane group. This net is invariant under any movement of the point group.

Proof. Let A be the centre of a rotation s in the plane group. Consider a net, this time with origin at A, made by applying the same translations T_1, T_2. Any point Q of this net is $T(A)$, where T is a translation in the group. Then

$$s(Q) = sT(A) = sTs^{-1}(A), \text{ since } s^{-1}(A) = A.$$

But sTs^{-1} is a translation in the group, so Q is moved to another point of the net. Therefore the net as a whole is unchanged by the rotation s. It follows that the net with origin at P is unchanged by the rotation S.

Similarly if m is a reflexion in the plane group, let A be any point on the mirror line. Then

$$m(Q) = mT(A) = mTm^{-1}(A),$$

and because mTm^{-1} is a translation in the group, Q is moved to another point of the net and the net as a whole is unchanged by m. It follows that the net at P is unchanged by the corresponding reflexion M.

If g is a glide reflexion in the plane group,

$$g = nt = tn,$$

where n is a reflexion and t a translation, neither of them being necessarily members of the group. If A' is chosen on the glide line, we cannot say that the net with origin at A' is unchanged by g. But gTg^{-1} is a translation in the group and

$$gTg^{-1} = ntTt^{-1}n^{-1} = nTn^{-1},$$

since translations are commutative; so nTn^{-1} is a member and

$$n(Q) = nT(A') = nTn^{-1}(A'),$$

another point of the net. Hence the net is unchanged by n and the net at P is unchanged by the corresponding M.

From this it appears that the rotations and reflexions S and M are symmetry movements of the net at P as well as of the point group. This enables us to find all the possible plane groups from the five nets and the ten point groups.

We choose a net, and then a point group consisting of movements that are symmetry movements of the net. But in working back to the plane groups we must bear in mind that a reflexion in the point group may correspond either to reflexions or glide reflexions in the plane group. Thus, for example, the point group $1M$ leads to the plane groups $p1m$ and $p1g$. The complete list is shown in Table 18.1.

Table 18.1

Net	Symmetries of net	Symmetries of point group	Symmetries in common	Plane groups
Parallelogram	2	1	1	$p1$
		2	2	$p2$
Rectangular	$2MM$	$1M$	$1M$	$p1m, p1g$
		$2MM$	$2MM$	$p2mm, p2mg, p2gg$
Centred rectangular	$2MM$	$1M$	$1M$	$c1m$
		$2MM$	$2MM$	$c2mm$
Square	$4MM$	4	4	$p4$
		$4MM$	$4MM$	$p4mm, p4gm$
Hexagonal	$6MM$	3	3	$p3$
		$3M$	$3M$	$p3m1, p31m$
		6	6	$p6$
		$6MM$	$6MM$	$p6mm$

We must first be sure that these groups are all different. That $p1m$ and $c1m$ are distinct is seen more clearly if the intermediate reflexion or glide lines are considered: there are reflexions for $p1m$ and glides for $c1m$. (See the lower line of symbols in Fig. 18.05.) The same difference occurs with $p2mm$ and $c2mm$. The two groups $p3m1$ and $p31m$ are different because in $p3m1$ the combination of a reflexion and a minimum translation may be another reflexion, but in *p*31*m* this is not so (Fig. 18.03).

Further, we must consider whether the groups listed above are the only ones. Not every *M* in the point groups can be replaced by *g* to make a new group. As explained on p. 19, it cannot be done with a centred net. This applies not only to the centred rectangular net but also to the hexagonal one, since that can be regarded as a special case of the centred rectangle. With tetrad rotation we cannot have *p*4*gg*, because glide lines in one direction then imply simple reflexions in perpendicular lines or in lines at 45°. This is shown in Fig. 18.04. where in (*a*) the glide lines run through the centres of rotation and in (*b*) they do not. (As usual, these are shown by broken lines, with continuous lines for the consequent reflexions.)

The 17 two-dimensional plane groups were shown together in Fig. 4.36 (p. 25). They are shown again in Fig. 18.05 in a diagrammatic form which was used in Part I for three-dimensional groups and will be needed again in later chapters. Each diagram represents a unit cell, with dots to show a set of equivalent positions. The number of movements in the group (other than the translations and consequent combinations) is the same as the number of dots. The red dots are reflected versions of the ones in black, so that a movement from black to red, or vice versa, is always an opposite movement (reflexion or glide); whereas black to black or red to red is a direct movement (rotation or translation). The origin (at the top left-hand corner of each diagram) is a centre of minimum rotation.

Fig. 18.03

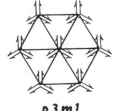

p 3 *m* 1

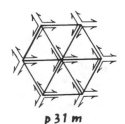

p 31 *m*

Fig. 18.04

(*a*)

(*b*)

Fig. 18.05

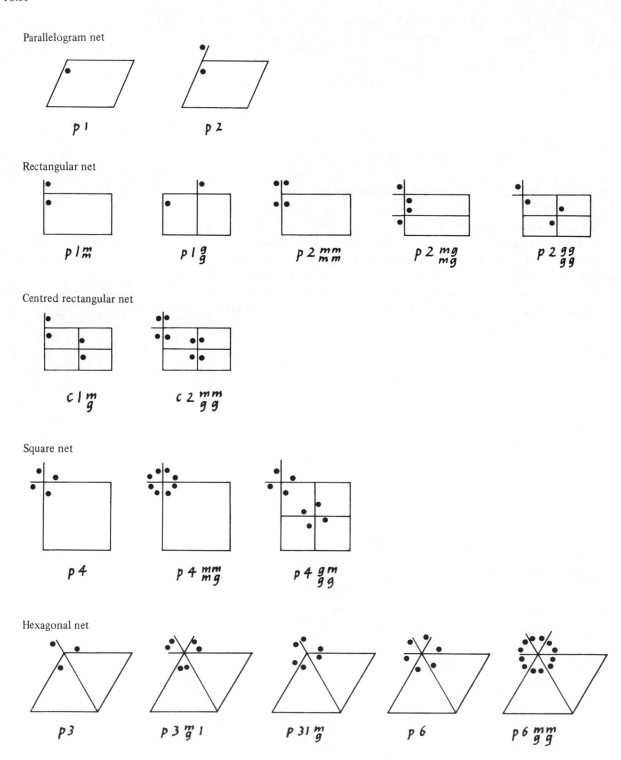

Parallelogram net

$p\,1$ $p\,2$

Rectangular net

$p\,1\frac{m}{m}$ $p\,1\frac{g}{g}$ $p\,2\frac{mm}{mm}$ $p\,2\frac{mg}{mg}$ $p\,2\frac{gg}{gg}$

Centred rectangular net

$c\,1\frac{m}{g}$ $c\,2\frac{mm}{gg}$

Square net

$p\,4$ $p\,4\frac{mm}{mg}$ $p\,4\frac{gm}{gg}$

Hexagonal net

$p\,3$ $p\,3\frac{m}{g}1$ $p\,31\frac{m}{g}$ $p\,6$ $p\,6\frac{mm}{gg}$

19

Movements in three dimensions

In three dimensions isometries can still be direct or opposite. If ABC and $A'B'C'$ are congruent triangles and D is a point not in the plane of ABC, there are two possible positions for D' such that $D'A' = DA$, $D'B' = DB$ and $D'C' = DC$, i.e. such that the tetrahedrons $ABCD$ and $A'B'C'D'$ are congruent. If P and P' are the feet of the perpendiculars drawn from D and D' to the planes of ABC and $A'B'C'$ respectively, the circuits ABC and $A'B'C'$ may be clockwise or anticlockwise in relation to the directions of PD and $P'D'$. If they are in the same sense the isometry is *direct*; if in opposite senses it is *opposite*, and the figures are then said to be *enantiomorphic*.

An isometry in three dimensions is determined by three non-collinear points and their corresponding points, together with a statement either that it is direct or that it is opposite.

Movements with one point fixed

With one point fixed, an isometry, direct or opposite, is determined by the movement of two other points and these may be taken at equal distances from the fixed point O. They will then be on a sphere whose centre is O and the movement will be completely represented by a movement on the surface of the sphere.

Direct movements

Any direct isometry can be effected by a rotation. For suppose that A and B on the sphere move to A' and B' and the great circles bisecting AA' and BB' at right angles meet at C (Fig. 19.01) and also at D (not shown). Then $CA = CA'$ and $CB = CB'$ and hence, by congruent triangles, angle ACA' = angle BCB'. Therefore a rotation about the diameter CD will bring A to A' and B to B'. Moreover it is easily proved by congruent triangles that the same rotation will bring any other point P to a point P' such that triangles ABP and $A'B'P'$ are congruent.

It follows that any two rotations about diameters of the sphere combine to give a single rotation; but, as will be shown in the next chapter, the angles of rotation are not additive.

Opposite movements

One opposite movement is inversion in the centre O, any point P on the sphere being replaced by the diametrically opposite point P'. To see that the isometry is an opposite one, consider two points P, Q and their inverses P', Q' (Fig. 19.02). The movements PQ and $P'Q'$ on the sphere are movements in the same sense along a great circle, but any point R to one side of the circle will have its inverse R' on the other side. In the figure the turn from RP to RQ is a left-

Fig. 19.01

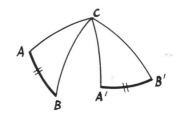

Fig. 19.02

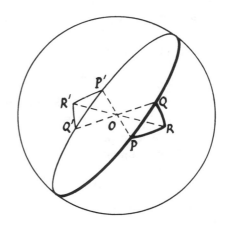

handed turn about the radius OR, while that from $R'P'$ to $R'Q'$ is right-handed about OR'. Inversion is thus an opposite movement.

Any opposite movement can be effected by a rotation followed by inversion. For let B be an opposite movement and let Z denote inversion. Then $ZB (= A)$ is a direct movement, and $B = ZZB = ZA$. On the sphere, A is either the identity or a rotation. Hence B is either simple inversion or the product of inversion and a rotation, which may be called *rotatory inversion*. Moreover the product is commutative: for the same rotation A that brings P to Q will also bring P' to Q', and the movement AZ takes P to P' and thence to Q', while ZA takes P to Q and so to Q'. So $ZA = AZ$. Furthermore the combination of Z with a rotatory inversion AZ is commutative: for $Z(AZ) = Z(ZA) = ZZA = A$, and $(AZ)Z = AZZ = A$.

Inversion is reflexion in a point. It is equivalent to reflexion in any plane through the point combined with a half-turn about an axis through the point perpendicular to that plane. In Fig. 19.03, P' is the inverse of P in the point O and P'' is both the reflexion of P in the plane and the result of giving P' a half-turn about the axis l. This also shows that reflexion is a special case of rotatory inversion.

Rotatory reflexion
This is reflexion in a plane combined with rotation about an axis perpendicular to that plane. It is an opposite movement and can therefore be expressed as a rotatory inversion. In Fig. 19.04 P is reflected to P' and then rotated through an angle a to Q'. The movement could equally well be effected by rotation through $(\pi + a)$ to Q, followed by inversion to Q'. Thus rotatory reflexion and rotatory inversion are two ways of describing the same movement. (In crystallography rotatory inversion is now preferred.)

Movements in general
Rotations about parallel axes combine in the same way as rotations in a plane (p. 101), i.e. they give a rotation about a third parallel axis, the angles of rotation being additive. A rotation and a translation perpendicular to the axis of rotation also combine as in two dimensions (p. 101), i.e. they give an equal rotation about a parallel axis.

A rotation and a translation along its axis form a *screw rotation*. If the angle of rotation is $2\pi/n$, where n is an integer, n such movements are equivalent to a translation.

> *Any direct isometry in three dimensions can be effected by a translation or a rotation or a screw rotation.*

Suppose the isometry takes three points A, B, C to positions A', B', C'. Let T be the translation that brings A to A'. If it also brings B to B' and C to C' the isometry is effected by the translation T. If not, the movement can be completed by a rotation S about an axis l through A'. If T is perpendicular to l the combined movement ST is a rotation about an axis l' parallel to l. If not (Fig. 19.05), let it be resolved into T_1 perpendicular to l and T_2 parallel to l. Then ST_1 is a rotation about an axis parallel to l and this with T_2 makes a screw rotation.

> *Any opposite isometry in three dimensions can be effected by an inversion or a reflexion or a glide reflexion or a rotatory reflexion.*

Let an inversion be chosen, centre O, to bring any point P to the corresponding

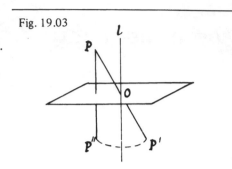

Fig. 19.03

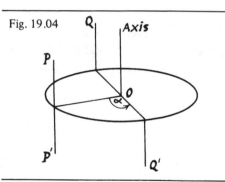

Fig. 19.04

point P' (Fig. 19.06, positions 1 and 3). Then the movement can if necessary be completed by a rotation about an axis l through P' (positions 3 to 4). The inversion can be replaced by a reflexion in a plane through O perpendicular to l (1 to 2) and a half-turn about an axis l' through O parallel to l (2 to 3). Then either the rotations about l and l' cancel out (the axes coinciding) and the whole movement is a reflexion; or they combine to give a translation, and then it is a glide reflexion; or they are together equivalent to a rotation about a parallel axis l'', and it is a rotatory reflexion.

Fig. 19.05

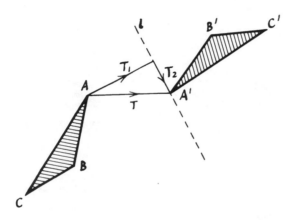

Fig. 19.06

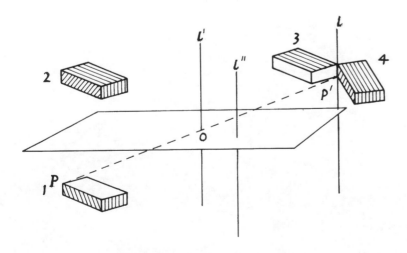

There are thus only the following movements to be considered:
    Direct movements:
    translations,
    rotations,
    screw rotations.
    Opposite movements:
    inversions,
    reflexions
    glide reflexions,
    rotatory reflexions (or rotatory inversions).

20

Point groups in three dimensions

A *three-dimensional point group* is a group of movements in three dimensions all of which leave one point fixed. It can be completely represented by the corresponding group of movements on the surface of a sphere, the centre of the sphere being the fixed point.

Direct movements

We first consider groups that contain only direct movements with, of course, the identity. These are necessarily rotations, the angles of rotation about any axis being multiples of $2\pi/n$, where n is some positive integer (as proved on p. 106).

One axis of rotation

With one axis of rotation only we have a series of groups some of which are represented in Fig. 20.01. The value of n is used as a symbol for the group, 1 representing a complete turn, equivalent to no rotation at all. In these diagrams the circle, as before, represents one hemisphere in stereographic projection and the dots represent successive positions of a single point. The centre of the circle represents the axis of rotation. It is convenient to think of the dots as representing points in the 'northern' hemisphere, projected from the 'south pole' onto the 'equatorial' plane, this plane being shown as viewed from the 'north' side. In later diagrams of this sort rings will be used to represent points in the 'southern' hemisphere, as projected onto the same plane from the 'north' pole. (See Fig. 5.23, p. 33.)

Fig. 20.02

More than one axis of rotation

We next consider the possibility of more than one rotation axis. Any two rotation axes imply a third, as may be proved as follows:

Let O be the centre of a sphere and let ABC be a spherical triangle having angles α, β, γ at A, B, C respectively (Fig. 20.02). Then if A, B, C is the clockwise order, anticlockwise rotations of $2\alpha, 2\beta, 2\gamma$ about A, B, C respectively

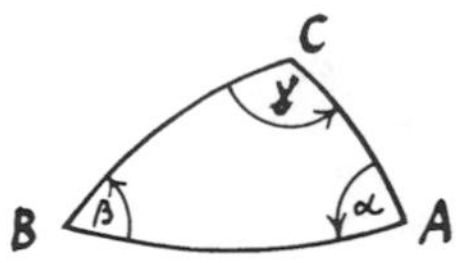

Fig. 20.01

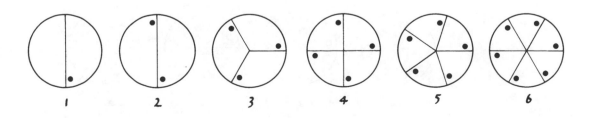

leave the figure unchanged: for the first rotation is equivalent to successive reflexions in the planes OAC, OAB; the next to reflexions in OBA, OBC; and the third to reflexions in OCB, OCA. But these reflexions cancel out in pairs. This is Sylvester's Theorem, as mentioned on p. 100. It follows that anticlockwise rotations of 2α about A and 2β about B combine in that order to give a clockwise rotation of 2γ about C. This is Euler's Construction, as mentioned on the same page; but $\alpha + \beta + \gamma$ is now greater than π, so the angles of rotation are no longer additive.

If now A and B are centres of rotation as near together as possible and the rotations 2α and 2β are minimum rotations $2\pi/p$ and $2\pi/q$ about A and B respectively, the rotation 2γ about C will also be a minimum rotation (for otherwise there would be another centre on the great circle between A and B, contrary to hypothesis). Calling the angle of this third rotation $2\pi/r$, the angles of the triangle are π/p, π/q, π/r, and it follows that $\frac{1}{p} + \frac{1}{q} + \frac{1}{r} > 1$.

The only solutions for this inequality are:

    2, 2, n, where n is any integer greater than 1;
    2, 3, 3;
    2, 3, 4;
    2, 3, 5.

With the solutions 2, 2, n, two angles of the triangle, say A and B, are right angles, so we may conveniently take A and B as points on the 'equator', with C at the 'north pole'. Some of the resulting groups are shown in Fig. 20.03. As before, the first numeral in the symbol refers to rotation about the 'polar' or z-axis, the second to a set of axes in the 'equatorial' or xy-plane related to one another by the first rotation, and the third, if any, to another such set similarly related.

The remaining solution sets of the inequality give the rotation groups of the regular solids: 2, 3, 3 for the tetrahedron, 2, 3, 4 for the cube or octahedron, 2, 3, 5 for the dodecahedron or icosahedron.

For the 2, 3, 3 set, the angles of the triangles are $\pi/2$, $\pi/3$, $\pi/3$, and the area is equal to the spherical excess, $\pi/6$.* As the whole area of the sphere is 4π, there will be 24 such triangles. The 24 angles of $\pi/2$ will meet four at a time in 6 vertices, and the 48 angles of $\pi/3$ will meet six at a time in 8 vertices. This gives 3 diad axes and 4 triad. As each set must be symmetrically arranged, the three diads will be mutually perpendicular and can be taken as coordinate axes, with the four triads running from the origin to the points $(1, 1, 1)$, $(1, -1, -1)$,

---

*Area of a spherical triangle $ABC = \angle A + \angle B + \angle C - \pi$.

Fig. 20.03

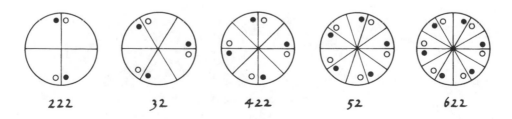

222   32   422   52   622

$(-1, 1, -1)$, $(-1, -1, 1)$. Fig. 20.04(a) shows this in the positive octant and Fig. 20.04(b) shows the whole sphere in stereographic projection. This is the rotation group of the rectangular tetrahedron, as shown in Fig. 20.04(c). The group contains 12 members.

There are several methods by which the number of members may be counted:

(i)   the number of movements (including the identity) is the same as the number of points in the stereogram (Fig. 20.04(b)).

(ii) the three diad axes provide 3 possible movements, apart from the identity; and the four triads provide 4×2, i.e. 8, further movements, making 12 in all.

(iii) the number of members is half the number of triangles.

The last statement may be proved as follows.

Let A_1 be a p-vertex (i.e. a centre of p-fold rotation, where $2p$ triangles meet) and let $S_1, S_2, \ldots, S_p$ be the p movements of the group that leave A_1 fixed, S_1 being the identity. Suppose that other members of the group take A_1 to $A_2, A_3, \ldots, A_n$. Let R_2 be one of the movements that take A_1 to A_2. Then $R_2 S_1 (= R_2), R_2 S_2, \ldots, R_2 S_n$ are all movements of the group and all of them take A_1 to A_2. If $R_3, \ldots, R_n$ are similarly defined, we have the following movements, np in number, all members of the group:

$$
\begin{array}{ccccc}
S_1 & S_2 & . \quad . \quad . & S_p \\
R_2 S_1 & R_2 S_2 & . \quad . \quad . & R_2 S_p \\
. & & . \quad . \quad . & . \\
R_n S_1 & R_n S_2 & . \quad . \quad . & R_n S_p
\end{array}
$$

These movements are all different, since those in different rows take A_1 to different positions, and within a row the Ss are all different. Moreover every movement in the group is included, for if a movement X takes A_1 to A_l, $R_l^{-1} X$ is a movement of the group leaving A fixed and therefore a member of the top row, S_k say. Then $X = R_l S_k$ and is a member of the row containing

Fig. 20.04

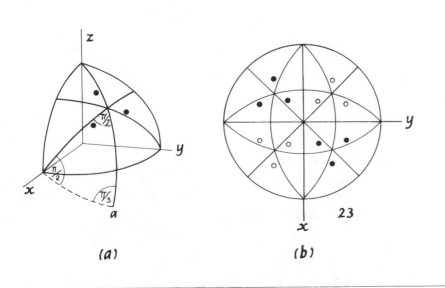

(a)   (b)

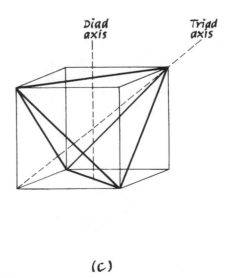

(c)

R_lS_1. It follows that the total number of movements in the group, N, is equal to np.

Now suppose that there are n_1 p-vertices, n_2 q-vertices and n_3 r-vertices. Then $N = n_1p = n_2q = n_3r$. The total number of angles at these vertices is $2pn_1 + 2qn_2 + 2rn_3$ (since $2p$ triangles meet at each centre of p-fold rotation), i.e. $6N$, and as each triangle provides three angles, the number of triangles is $2N$.

In this and related types (e.g. the next one) the symbols have a new meaning. The first numeral refers to rotations about the x-, y- and z-axes; the second to the four axes through $(1, 1, 1)$, etc.; and the third, if any, to the six axes through $(1, 1, 0)$, $(1, -1, 0)$, etc. For the tetrahedral rotation group the symbol is 23.

For the 2, 3, 4 set of values the angles are $\pi/2$, $\pi/3$, $\pi/4$, and the area of the triangle is $\pi/12$. There are therefore 48 triangles, the angles of $\pi/2$ meeting four at a time in 12 vertices, those of $\pi/3$ six at a time in 8 vertices, and those of $\pi/4$ eight at a time in 6 vertices. This gives 6 diad axes, 4 triad and 3 tetrad. Taking the three tetrads as axes of coordinates, the triads will be placed as before and the diads will bisect the angles formed by the tetrads. The symbol is therefore 432 and the group contains 24 members. This is shown in Fig. 20.05(a), (b). The positive octant has the same appearance as before, but there are now points in all eight octants instead of only four of them. This is the rotation group of the cube and octahedron (Fig. 20.05(c), (d)).

For the 2, 3, 5 set of values the angles are $\pi/2$, $\pi/3$ and $\pi/5$, and the area of the triangle is $\pi/30$. There are 120 triangles, the angles of $\pi/2$ meeting four at a time in 30 vertices, those of $\pi/3$ six at a time in 20 vertices, and those of $\pi/5$ ten at a time in 12 vertices. This gives 15 diad axes, 10 triad and 6 pentad. This is the rotation group of the icosahedron and dodecahedron. The symbol is 532, the pentad axes being related by the triad and diad rotations, and so on. The group contains 60 members. There are no convenient rectangular coordinate axes and the stereographic form of diagram is less suitable than in other cases.

Fig. 20.05

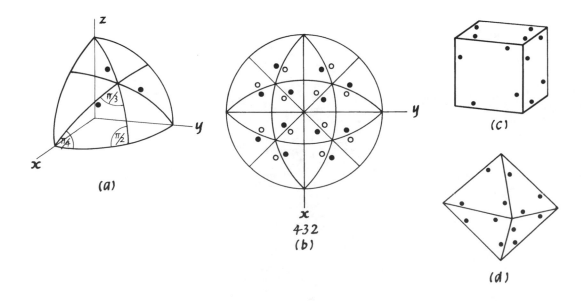

(a)

432
(b)

(c)

(d)

Fig. 20.06 shows an icosahedron and a dodecahedron with dots on the front faces to represent corresponding positions of a point.

Direct and opposite movements

We next consider groups containing opposite movements. In any such group the direct movements, including the identity, form a sub-group, for any two direct movements combine to give a direct movement. Furthermore the numbers of direct and opposite movements are equal. To prove this, suppose that there are n direct movements, $A_1, A_2, \ldots, A_n$, of which $A_1 = I$ (the identity), and m opposite movements, $B_1, B_2, \ldots, B_m$. Then B_1 combines with each of the A-movements to give a B-movement; and the B-movements so formed are all different (for if $B_1 A_r = B_1 A_s$, then $B_1^{-1} B_1 A_r = B_1^{-1} B_1 A_s$, i.e. $A_r = A_s$). Therefore $m \geqslant n$. Likewise B_1 combines with each of the B-movements (including itself) to give m different A-movements: therefore $n \geqslant m$. It follows that $m = n$.

Groups containing inversion

Groups containing opposite movements are of two kinds: those that contain inversion (denoted by Z) and those that do not. For the first kind, since Z is a member, so also are $ZA_2, \ldots, ZA_n$, and these are all opposite movements. The whole group therefore consists of the sub-group of direct movements and their combinations with inversion.

Each of the groups described above gives rise to a group of this sort. We thus have two more infinite sets of groups and three more polyhedral groups. In Fig. 20.07 we show some of the new groups below the rotational groups from which they are derived. (This time we show only those groups of interest to crystallographers, i.e. those that can occur in patterns that repeat in more than one direction). Opposite movements are shown by a change of colour, from black to red or from red to black. The full symbols are shown, with shortened forms in parentheses.

All these derived groups of the first kind have rotation *and* rotatory inversion, but those with even-numbered rotations about the z-axis have reflexion in the xy-plane as well. (This is a consequence of the fact that a half-turn combines with inversion to give a reflexion.) Thus the symbols in the second row of the figure are $2/m$, $4/m$ and $6/m$ for the even-numbered groups and $\bar{1}$, $\bar{3}$ for the odd-numbered. ($\bar{3}$ includes 3 and $2/m$ includes $\bar{2}$, etc.).

One non-crystallographic group of this kind deserves special mention: it is the one derived by adding inversion to the icosahedral rotation group 532. As might be expected from the last-mentioned considerations, it contains pentad

Fig. 20.06

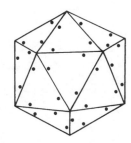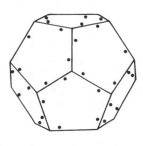

Fig. 20.07

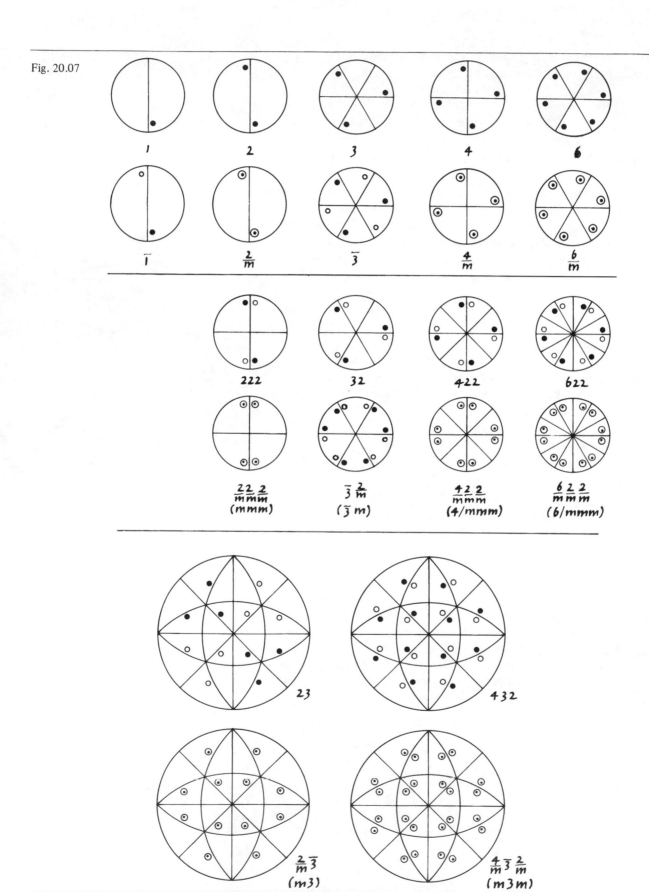

and triad rotation and rotatory inversion, with reflexion planes (15 of them) perpendicular to the diad axes. The symbol is $\overline{5}\,\overline{3}\frac{2}{m}$. It is the symmetry group of the regular icosahedron and dodecahedron and contains 120 members (Fig. 20.08).

Groups not containing inversion

For the second kind of point group containing opposite movements (i.e. those that do not contain Z), suppose that the group consists of direct movements $A_1, A_2, \ldots, A_n$, where $A_1 = I$, and opposite movements $B_1, B_2, \ldots, B_n$. Then $ZB_1, ZB_2, \ldots, ZB_n$ are all direct movements and $A_1, A_2, \ldots, A_n$, $ZB_1, ZB_2, \ldots, ZB_n$ form a group. To prove this we note that

(i) the identity (A_1) is a member;

(ii) each ZB, as well as each A, has its inverse (for if B_r has B_s as its inverse in the original group, ZB_r and ZB_s will be inverses in the new group);

(iii) symmetry movements are always associative;

(iv) the combination of a ZB with an A is a ZB (for in the original group $B_r A_s$ is a B, and therefore in the new group $ZB_r A_s$ is a ZB); and the combination of two ZBs is an A (for $ZB_r ZB_s = B_r ZZB_s = B_r B_s$, and $B_r B_s$ is an A).

So for a point group of this kind we must look first for a rotation group having a sub-group of half its members; we must then replace the members not in the sub-group (the coset) by their combinations with Z. For example, the group 6 contains the sub-group 3, so we can form a new group as shown in Fig. 20.09. The new group could be described as $3/m$, but equally well as $\overline{6}$. The latter is preferred, for reasons concerned with space groups.

A new group can be derived in this way from each of the groups 2, 4, 6, 222, 32 and 432; and two new groups each from 422 and 622. These ten new groups and those from which they are derived are shown in Fig. 20.10.

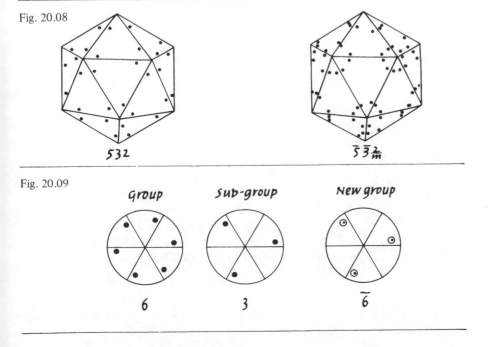

Fig. 20.08

532          $\overline{5}\,\overline{3}\frac{2}{m}$

Fig. 20.09

group   sub-group   New group

6   3 $\overline{6}$

Fig. 20.10

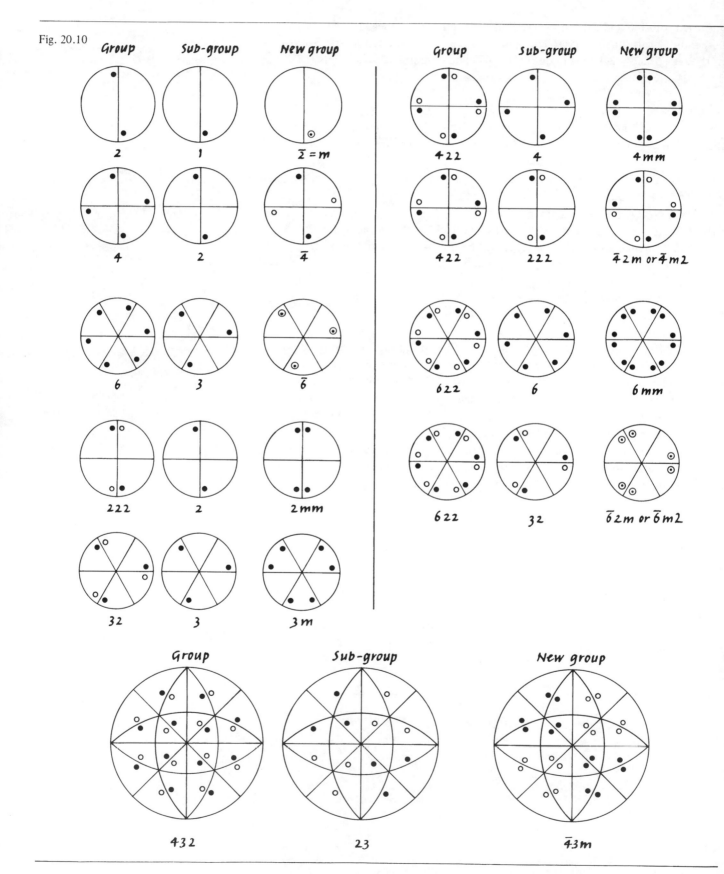

If we restrict ourselves to the 'crystallographic' rotations 1, 2, 3, 4 and 6, the ten groups shown in Fig. 20.10 are the only ones of their kind. With the 22 groups shown in Fig. 20.07 this makes 32 groups in all. For future reference it is convenient to list them (Table 20.1) in the order used in Fig. 5.25 (p. 34).

Table 20.1

1	2	3	4	6
$1m$	$2mm$	$3m$	$4mm$	$6mm$
m	$2/m$		$4/m$	$6/m$
$m2m$	$\dfrac{2}{m}\dfrac{2}{m}\dfrac{2}{m}$		$\dfrac{4}{m}\dfrac{2}{m}\dfrac{2}{m}$	$\dfrac{6}{m}\dfrac{2}{m}\dfrac{2}{m}$
12	222	32	422	622
$\bar{1}$		$\bar{3}$	$\bar{4}$	$\bar{6}$
$\bar{1}\dfrac{2}{m}$		$\bar{3}\dfrac{2}{m}$	$\bar{4}2m$	$\bar{6}2m$

and the polyhedral groups

$$\bar{4}3m \qquad 23 \qquad \frac{4}{m}\bar{3}\frac{2}{m} \qquad 432 \qquad \frac{2}{m}\bar{3}$$

($\frac{4}{m}\bar{3}\frac{2}{m}$ is abbreviated to $m3m$, $\frac{2}{m}\bar{3}$ to $m3$)

The list contains 36 members because it includes the four 'second settings', as explained in Chapter 5, namely:

$$1m, \quad m2m, \quad 12, \quad \bar{1}\tfrac{2}{m},$$

which are the same groups as

$$m, \quad 2mm, \quad 2, \quad \tfrac{2}{m},$$

though differently orientated. These 'second settings' are of use in the enumeration of line groups and plane groups, where one of the three axes is 'unique'. The four blanks in the table correspond to duplicate types. $3/m$ would be the same as $\bar{6}$, $\frac{3}{m}2m$ the same as $\bar{6}2m$, $\bar{2}$ the same as m, and $\bar{2}2m$ the same as $m2m$.

21

Line groups in three dimensions

A *line group* has already been defined as a group of movements that includes translations in one and only one direction. In two dimensions we found that it could include also a very limited number of rotations, reflexions and glide reflexions. In three dimensions a greater variety of these becomes possible and there may also be inversions, rotatory inversions and screw rotations. A *line ornament* (or *rod pattern*) is a pattern that remains invariant under such a group. As with friezes it can be divided into cells, the cells being related by the sub-group of translations T^r (r integral).

The direction of T will be described as *longitudinal* and will be taken as the z-direction. With an origin O, T determines a row of points on which the pattern is built up. A glide or screw, if repeated, gives a translation, so glides and screws can only be in the z-direction. The rotations and other movements are limited by their interactions with the translations. For let S be any rotation or screw rotation in the group. Then STS^{-1} is T rotated by S and is a translation in the group. It must be T or T^{-1}. Hence S can only be a rotation or screw about an axis in the z-direction or a half-turn rotation about a perpendicular (*transverse*) axis. Again, if M is a reflexion or glide reflexion in the group, MTM^{-1} is a translation restricted in the same way and M can only be a reflexion or glide in a plane parallel to T or a reflexion in one perpendicular to T.

In the longitudinal direction there can be only one axis of rotation or screw rotation, for two would imply a transverse translation. Any inversion centres must lie on it (otherwise there would be a parallel rotation axis) and any longitudinal reflexion plane must contain it (for the same reason). Similar considerations show that, if there is no such axis, it is still possible to choose a line in the direction of the translations to pass through any inversion centres and to lie in any longitudinal reflexion plane. We call this line the *axis* of the line group.

We now define a point group from which the line group can be built up. A plane is chosen perpendicular to the direction of the translations and the point where it is cut by the axis is the centre of the point group. The translations of the line group are mapped into the identity; half-turns about transverse axes are mapped into half-turns about parallel axes through the centre; reflexions in transverse planes are mapped into a single reflexion in the chosen plane; and inversions into a single inversion in the centre point. Longitudinal rotations and reflexions are mapped into themselves, screws into rotations and glides into reflexions. The new set of movements form a group, as can be seen from the fact that they combine in the same way as the movements from which they have been derived.

From this it is evident that a line group can always be formed from a point group by adding translations T', perhaps replacing rotations by screws and reflexions by glides, both in the direction of T.

The rotations and screws in the direction of T are now about a single axis, through the centre of the point group, and we take this as the z-axis. They are not confined to the 'crystallographic' numbers 1, 2, 3, 4 and 6, and thus there are an infinite number of line groups. For convenience in listing, however, we limit ourselves to the 'crystallographic' numbers and there are then 75 geometrically distinct groups.*

*Schubnikov and Koptsik describe 31 types of 'two-sided bands' (listed also by Bell & Fletcher as 'strip ornaments in relief'). These might be described as 'ribbon ornaments', the 'ribbon' having a front and a back, so that the x- and y-axes are now distinguishable. They consist of the 22 line ornaments that have no more than diad rotation, with 9 repetitions obtained by interchanging the x- and y-axes. They correspond exactly to the 31 types of black-and-white line ornament illustrated in Fig. 27.07 (p. 199).

Table 21.1

Point group	Derived line groups (the prefix r is omitted)			
1	1			
2	2	2_1		
3	3	3_1 and 3_2		
4	4	4_1 and 4_3	4_2	
6	6	6_1 and 6_5	6_2 and 6_4	6_3
$1m$	$1m$	$1c$		
$2mm$	$2mm$	$2_1 mc$	$2cc$	
$3m$	$3m$	$3c$		
$4mm$	$4mm$	$4_2 mc$	$4cc$	
$6mm$	$6mm$	$6_3 mc$	$6cc$	
m	m			
$2/m$	$2/m$	$2_1/m$		
$4/m$	$4/m$	$4_2/m$		
$6/m$	$6/m$	$6_3/m$		
$m2m$	$m2m$	$m2c$		
$\dfrac{2}{m}\dfrac{2}{m}\dfrac{2}{m}$ $(2/mmm)$	$\dfrac{2}{m}\dfrac{2}{m}\dfrac{2}{m}$	$\dfrac{2_1}{m}\dfrac{2}{m}\dfrac{2}{c}$	$\dfrac{2}{m}\dfrac{2}{c}\dfrac{2}{c}$	
$\dfrac{4}{m}\dfrac{2}{m}\dfrac{2}{m}$ $(4/mmm)$	$\dfrac{4}{m}\dfrac{2}{m}\dfrac{2}{m}$	$\dfrac{4_2}{m}\dfrac{2}{m}\dfrac{2}{c}$	$\dfrac{4}{m}\dfrac{2}{c}\dfrac{2}{c}$	
$\dfrac{6}{m}\dfrac{2}{m}\dfrac{2}{m}$ $(6/mmm)$	$\dfrac{6}{m}\dfrac{2}{m}\dfrac{2}{m}$	$\dfrac{6_3}{m}\dfrac{2}{m}\dfrac{2}{c}$	$\dfrac{6}{m}\dfrac{2}{c}\dfrac{2}{c}$	
12	12			
222	222	$2_1 22$		
32	32	$3_1 2$ and $3_2 2$		
422	422	$4_1 22$ and $4_3 22$	$4_2 22$	
622	622	$6_1 22$ and $6_5 22$	$6_2 22$ and $6_4 22$	$6_3 22$
$\bar{1}$	$\bar{1}$			
$\bar{3}$	$\bar{3}$			
$\bar{4}$	$\bar{4}$			
$\bar{6}$	$\bar{6}$			
$\bar{1}\dfrac{2}{m}$	$\bar{1}\dfrac{2}{m}$	$\bar{1}\dfrac{2}{c}$		
$\bar{3}\dfrac{2}{m}$ $(\bar{3}m)$	$\bar{3}\dfrac{2}{m}$	$\bar{3}\dfrac{2}{c}$		
$\bar{4}m2$	$\bar{4}m2$	$\bar{4}c2$		
$\bar{6}m2$	$\bar{6}m2$	$\bar{6}c2$		

Fig. 21.01

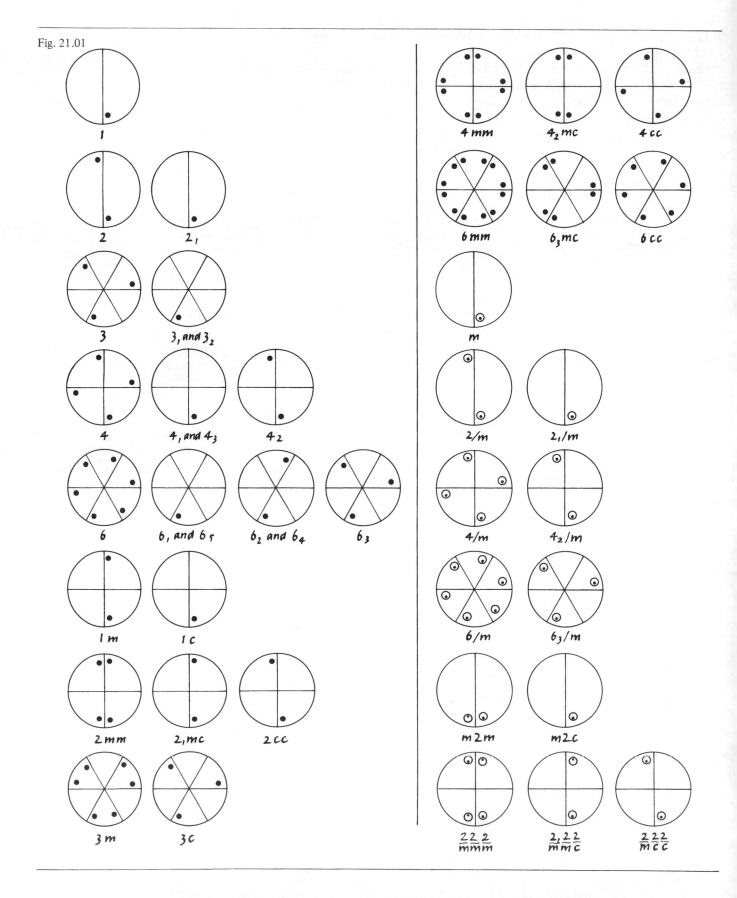

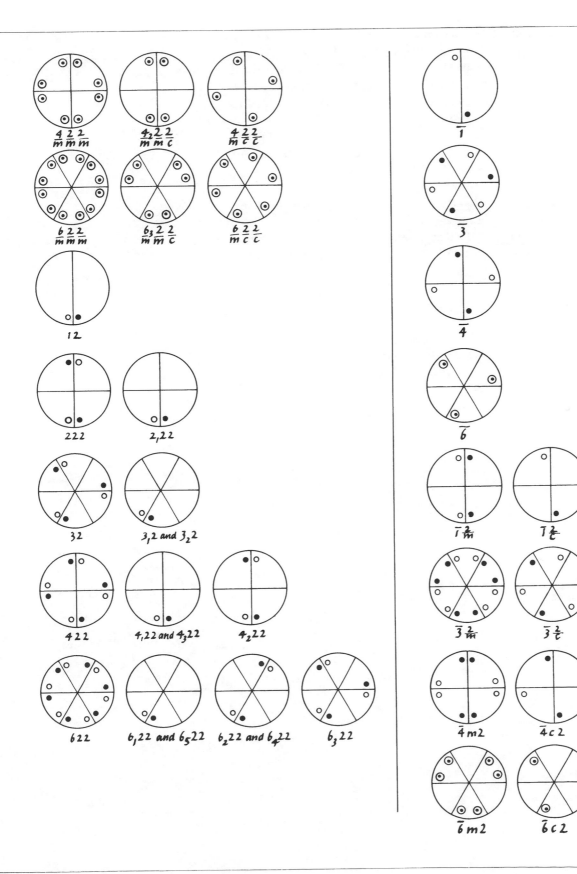

In Table 21.1 the line groups are shown alongside the point groups from which they are derived, enantiomorphic pairs being indicated by the word *and*. As before, the first position in each symbol is used to indicate rotation about the z-axis and reflexions in planes perpendicular to it; the second for a set of rotation axes in the xy-plane related to one another by the z-rotation, and for reflexions in planes perpendicular to those axes; and the third for a similarly related set of axes bisecting the angles formed by the previous set, and planes perpendicular to them. Glides in the z-direction are indicated by the letter c (to conform to the notation to be used in later chapters).

In the diagrams (Fig. 21.01) the same order of point groups is followed. The first diagram in each row shows the point group and also represents the line group obtained by simply adding the translations. Groups containing glides and screws are represented alongside, the diagram showing only those parts of the point group pattern that remain after other parts have been shifted in the z-direction.

22

Plane groups in three dimensions

In three dimensions as in two, a plane group is a group of movements that includes translations T_1, T_2 in different directions, with their combinations $T_1{}^p T_2{}^q$ (p, q integral) and no other translations. As in two dimensions, a plane group consists of movements that are symmetry movements both of a net and a point group. But in three dimensions a net has more symmetry movements than in two, and it will be seen that the number of available point groups is increased to 31. A *plane ornament* (or *layer pattern*) is one that remains invariant under such a group.

For the z-axis we choose the direction perpendicular to T_1 and T_2. It is convenient to call this direction 'vertical' and any plane or line perpendicular to it 'horizontal'. As with line groups, there are certain limitations on the movements. If T is a translation in the group and S a rotation, STS^{-1} is a translation in the group and is in fact T rotated by S. Hence S can only be a rotation about a vertical axis or a half-turn about a horizontal one. Similarly, if M is a reflexion in the group, MTM^{-1} is a translation in the group and hence M can only be a reflexion in a vertical plane or a horizontal one. There cannot be more than one horizontal reflexion plane, otherwise there would be translations out of the plane of T_1 and T_2; and, for the same reason, any horizontal half-turn axes or centres of inversion must lie in a plane, the same as the horizontal reflexion plane, if there is one.

As in two dimensions, we can set up a point group and a net, the centre of the point group (which is also the origin of the net) being in the plane mentioned above. Of the 32 three-dimensional point groups we can use all except the five polyhedral groups. Moreover the z-axis is now distinguishable from the other two, so the four 'second settings' may be used, making 31 groups in all.

On this basis we enumerate the plane groups, taking each of the five nets in turn, with the point groups whose symmetries they include.

Parallelogram net

Symmetries of the net are $2/m$, i.e. a half-turn about the z-axis and reflexion in the plane $z = 0$. There are of course further half-turn axes at each point of the net and at intermediate points. For glides in the xy-plane there is no distinction between the x- and y-directions; nor would any new type be found by taking glides diagonally, or in two directions at once. We therefore take glides in the y-direction only and indicate them by the letter b. The prefix p is used to indicate a primitive, as distinct from a centred, net. Table 22.1, and the diagrams in Fig. 22.01, show all possible plane groups with this net.

Table 22.1

Point group	Plane groups
1	$p1$
2	$p2$
m	pm, pb
$2/m$	$p2/m, p2/b$
$\bar{1}$	$p\bar{1}$

Fig. 22.01

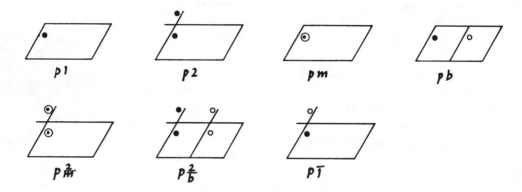

Rectangular net

The symmetries of the net are $\frac{2}{m}\frac{2}{m}\frac{2}{m}$. The three positions in the symbol refer to the axes in the order zxy, with reflexions in the planes $z = 0$, $x = 0$, $y = 0$. We shall call these planes the Z-, X- and Y-planes respectively. Glide reflexions in the Z-plane can be in any of three directions, the x-axis, the y-axis and diagonally. These will be denoted by a, b and n respectively. There may also be glide reflexions in the X- and Y-planes: in the X-plane a b-glide, and in the Y-plane an a-glide. The only possible screws are half-turn screws about axes in the Z-plane. There are 32 groups as listed in Table 22.2.

Table 22.2

Point group	Plane groups			
$1m$	$p1m$	$p1b$		
$2mm$	$p2mm$	$p2bm$	$p2ba$	
$m2m$	$pm2m$	$pm2_1a$	$pa2_1m$	$pa2a$
	$pn2_1m$	$pn2a$	$pb2m$	$pb2_1a$
$\dfrac{2\ 2\ 2}{m\,m\,m}$	$p\dfrac{2}{m}\dfrac{2}{m}\dfrac{2}{m}$	$p\dfrac{2}{m}\dfrac{2}{b}\dfrac{2_1}{m}$	$p\dfrac{2}{m}\dfrac{2_1}{b}\dfrac{2_1}{a}$	
	$p\dfrac{2}{b}\dfrac{2}{m}\dfrac{2_1}{m}$	$p\dfrac{2}{b}\dfrac{2_1}{m}\dfrac{2_1}{a}$	$p\dfrac{2}{b}\dfrac{2}{b}\dfrac{2}{m}$	$p\dfrac{2}{b}\dfrac{2_1}{b}\dfrac{2}{a}$
	$p\dfrac{2}{n}\dfrac{2_1}{m}\dfrac{2_1}{m}$	$p\dfrac{2}{n}\dfrac{2_1}{b}\dfrac{2}{m}$	$p\dfrac{2}{n}\dfrac{2}{b}\dfrac{2}{a}$	
12	$p12$	$p12_1$		
222	$p222$	$p222_1$	$p22_12_1$	
$\bar{1}\dfrac{2}{m}$	$p\bar{1}\dfrac{2}{m}$	$p\bar{1}\dfrac{2}{b}$	$p\bar{1}\dfrac{2_1}{m}$	$p\bar{1}\dfrac{2_1}{b}$

It is necessary to keep in mind that reflexions in intersecting planes imply a simple rotation about the line of intersection. Thus, for example, we can have $m2m$, but not $m2_1m$. If one of the reflexions is replaced by a glide the effect depends on the direction of the glide: if it is parallel to the line of intersection it changes the rotation to a screw (e.g. $a2_1m$), but if it is in a perpendicular direction it merely displaces the rotation axis by $\frac{1}{4}$ unit in the direction of the glide. Two glides parallel to the line of intersection have no effect on the rotation: thus we may have $a2a$ or $n2a$ (the n-glide having a component in the x-direction).

As it now appears that the rotations and screws are a consequence of the reflexions and glides, we find it convenient, in drawing the diagrams, to take the origin at the intersection of the reflexion or glide planes, whether that point happens to be on a rotation axis or not (Fig. 22.02). Each diagram shows a unit cell, with the origin at the top left-hand corner. A rotation axis is displaced from that point if there is a glide reflexion in the direction perpendicular to it in a plane parallel to it.

Fig. 22.02

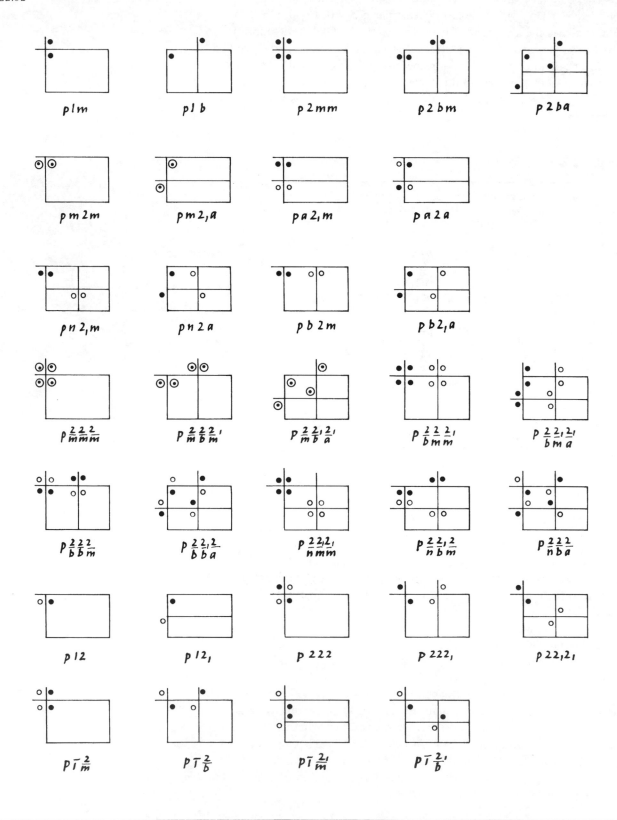

With the point group $m2m$, the x- and y-axes are distinguished by the diad rotation about the x-axis, but in the other point groups of this set there is no such distinction. Thus with the $2mm$ point group we do not need $p2ma$ as well as $p2bm$. With the group $\frac{2}{m}\frac{2}{m}\frac{2}{m}$, however, a b-glide in the first position provides a distinction for the subsequent choice.

Centred rectangular net

The symmetries are again $\frac{2}{m}\frac{2}{m}\frac{2}{m}$. There are no glides in 'vertical' planes, for they would produce simple reflexions. There may be glides in the Z-plane, but they must be in both x- and y-directions together. The letter g will be used to indicate this. The nine groups are shown in Table 22.3 and Fig. 22.03.

Table 22.3

Point group	Plane groups	
$1m$	$c1m$	
$2mm$	$c2mm$	
$m2m$	$cm2m$	$cg2m$
$\frac{2}{m}\frac{2}{m}\frac{2}{m}$	$c\frac{2}{m}\frac{2}{m}\frac{2}{m}$	$c\frac{2}{g}\frac{2}{m}\frac{2}{m}$
12	$c12$	
222	$c222$	
$\bar{1}\frac{2}{m}$	$c\bar{1}\frac{2}{m}$	

Fig. 22.03

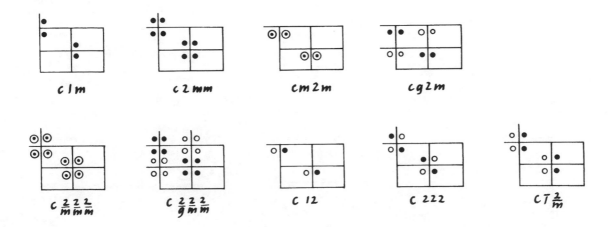

$c\,1\,m$ $c\,2\,mm$ $cm\,2\,m$ $cg\,2\,m$

$c\frac{2}{m}\frac{2}{m}\frac{2}{m}$ $c\frac{2}{g}\frac{2}{m}\frac{2}{m}$ $c\,12$ $c\,222$ $c\bar{1}\frac{2}{m}$

Square net

The symmetries are $\frac{4}{m}\frac{2}{m}\frac{2}{m}$, the second position in the symbol referring to both x- and y-axes, and to X- and Y-planes, the third to diagonal lines and planes. As the x- and y-axes have to be alike, the only possible glides are in both these directions (again denoted by g) or diagonally (denoted by n). We cannot have g in the first position, as that would lead to a smaller square; nor in the third position, since it would combine with a translation to give a simple reflexion in a parallel line (Fig. 22.04(a)). Again, we cannot have 2_1 in the third position, as the screw would combine with a translation to give a simple rotation about a parallel axis (Fig. 22.04(b)). There are 16 groups (Table 22.4 and Fig. 22.05). It is to be noted that, whereas point groups $\bar{4}m2$ and $\bar{4}2m$ are the same, they count separately here because they are differently orientated with respect to the translations. A similar difference occurred with the two-dimensional plane groups $p3m1$ and $p31m$.

In the diagrams for these groups (Fig. 22.05) we keep the origin on the tetrad axis, as the rotation would not be so clearly shown on the other system.

Table 22.4

Point group	Plane groups			
4	$p4$			
$4mm$	$p4mm$	$p4gm$		
$4/m$	$p4/m$	$p4/n$		
$\dfrac{4\ 2\ 2}{m\ m\ m}$	$p\dfrac{4}{m}\dfrac{2}{m}\dfrac{2}{m}$	$p\dfrac{4}{m}\dfrac{2_1}{g}\dfrac{2}{m}$	$p\dfrac{4}{n}\dfrac{2_1}{m}\dfrac{2}{m}$	$p\dfrac{4}{n}\dfrac{2}{g}\dfrac{2}{m}$
$\bar{4}$	$p\bar{4}$			
422	$p422$	$p42_12$		
$\bar{4}m2$	$p\bar{4}m2$	$p\bar{4}g2$	$p\bar{4}2m$	$p\bar{4}2_1m$

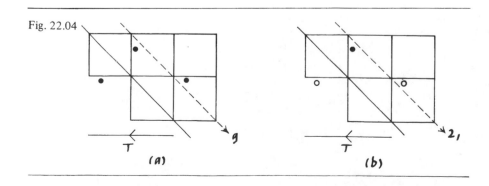

Fig. 22.04

(a)   (b)

Fig. 22.05

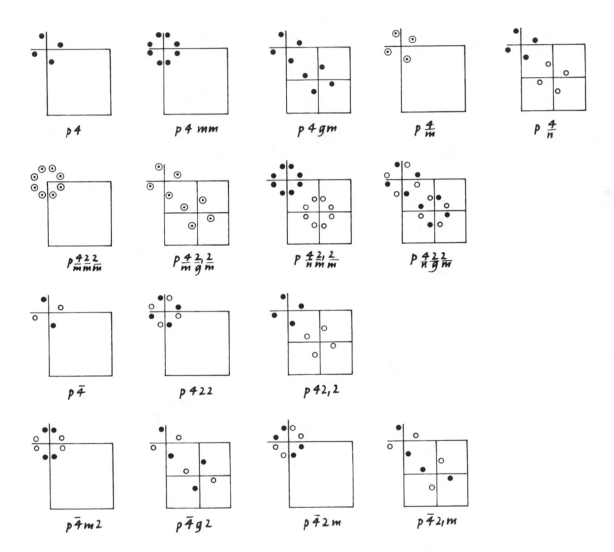

Hexagonal net

For this net it is convenient to have three axes, x, y and u, making angles of 120° with one another (Fig. 22.06). The symmetries are $\frac{6}{m}\frac{2}{m}\frac{2}{m}$, the second position referring to the directions of x, y and u, with planes perpendicular to them, the third to directions and planes bisecting the angles of 60° so formed. The net is, from one point of view, a centred rectangular one, and for that reason glides are not possible. There are again 16 groups (Table 22.5 and Fig. 22.07). This completes the list of the 80 plane groups in three dimensions.

Table 22.5

Fig. 22.06

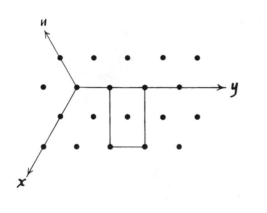

Point groups	Plane groups	
3	$p3$	
$3m$	$p3m1$	$p31m$
$\overline{3}$	$p\overline{3}$	
32	$p321$	$p312$
$\overline{3}\frac{2}{m}$	$p\overline{3}\frac{2}{m}1$	$p\overline{3}1\frac{2}{m}$
6	$p6$	
$6mm$	$p6mm$	
$6/m$	$p6/m$	
$\frac{6}{m}\frac{2}{m}\frac{2}{m}$	$p\frac{6}{m}\frac{2}{m}\frac{2}{m}$	
$\overline{6}$	$p\overline{6}$	
622	$p622$	
$\overline{6}m2$	$p\overline{6}m2$	$p\overline{6}2m$

Fig. 22.07

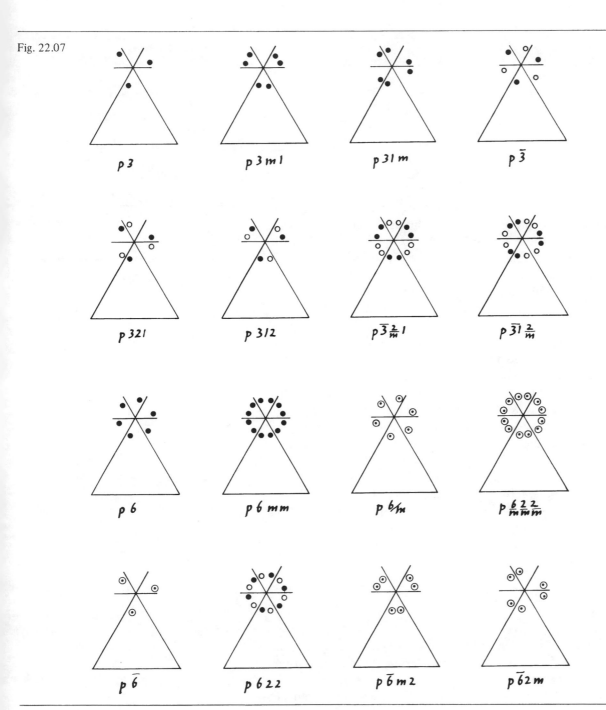

23

Lattices

A *lattice* is an array of points, not all in the same plane, determined by a point O and three translation vectors T_1, T_2, T_3. Any point of the lattice may be reached from O by the translation $T_1{}^p T_2{}^q T_3{}^r$, where p, q, r are integers. With O as origin and axes in the directions of T_1, T_2, T_3, the point may be described by the coordinates (p, q, r).

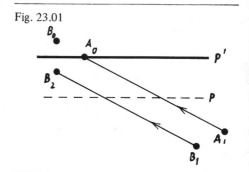

Fig. 23.01

If A_1 and A_2 are points of the lattice, $A_1 A_2$ is a translation vector which, applied to any lattice point A_3, would bring it to another lattice point: for if A_1, A_2 and A_3 are (p_1, q_1, r_1), (p_2, q_2, r_2) and (p_3, q_3, r_3), the new point has integral coordinates $(p_3 + p_2 - p_1, q_3 + q_2 - q_1, r_3 + r_2 - r_1)$. Thus any three lattice points determine a net, all of whose points are points of the lattice.

The origin and two of the translations determine a net, of one of the five types described in Chapter 17. The lattice may be built up from such a net by applying the third translation, thus forming a series of identical nets in parallel planes. We have to consider the types of lattice, as distinguished by their symmetries, and we seek in this chapter to prove that the 14 types described in Chapter 8 are the only ones.

Every lattice has inversion, in the lattice points and the mid-points of their joins, and in no other points. In terms of the unit cell, which is a parallelepiped, there is inversion in the vertices, the mid-points of the edges, the centres of the faces, and the body-centre. The other symmetry movements to be considered are:

> rotation about an axis,
> screw rotation,
> reflexion in a plane,
> glide reflexion,
> rotatory inversion.

We propose to show that the last four of these movements imply rotations and hence that we can find all possible lattices by considering all possible rotations. For this, four propositions are necessary:

> (i) *If a lattice has glide reflexion in any plane it also has simple reflexion in parallel planes.*

Suppose that A_0, B_0 are any two lattice points and that a glide reflexion G in a plane p carries them to A_1, B_1 (Fig. 23.01). Then $A_1 A_0$ is a translation of the lattice, i.e. it leaves the lattice as a whole unchanged. It carries B_1 to another lattice point B_2, and this is the reflexion of B_0 in a plane p' through A_0 parallel to p. Therefore the whole lattice is self-coincident under reflexion in p'.

> (ii) *If a lattice has screw rotation about any axis it also has the corresponding simple rotation about parallel axes.*

Suppose that S is a screw of angle $2\pi/n$ about an axis l and that it carries lattice points A_0, B_0 to A_1, B_1 (Fig. 23.02(*a*)). Then $A_1 A_0$ is a translation of the lattice and carries B_1 to another lattice point B_2, where B_0 and B_2 lie in a plane perpendicular to l and $A_0 B_2 = A_1 B_1 = A_0 B_0$. Projecting the figure onto a plane perpendicular to l, with A_0 projected to a_0, etc. (Fig. 23.02(b)), triangle $la_1 b_1$ is triangle $la_0 b_0$ turned through $2\pi/n$ about l, and hence angle $b_0 a_0 b_2 = 2\pi/n$. Thus a rotation of $2\pi/n$ about an axis l' drawn through A_0 parallel to l leaves the whole lattice unchanged.

    (iii) *If a lattice has reflexion in any plane it also has diad rotation about axes perpendicular to that plane.*

Let A_0, B_0 be any two lattice points and A_1, B_1 their images in the reflexion plane p (Fig. 23.03). Then $A_1 B_1$ is a translation of the lattice and takes B_0 to a point B_2 of the lattice, where A_0 and B_2 are in a plane parallel to p. But B_2 is the result of giving A_0 a half-turn about an axis through B_0 perpendicular to p. Therefore the whole lattice is self-coincident under such a movement.

    (iv) *If a lattice has rotatory inversion (other than simple inversion) about any axis it also has rotation about parallel axes through the lattice points.*

With O as centre of inversion, let A_0, B_0 be any two lattice points and suppose that the rotatory inversion, applied twice, takes them to A_2, B_2. Then $A_2 B_2$ is a translation of the lattice and it would take A_0 to another lattice point C. Fig. 23.04 shows these points projected onto a plane through O perpendicular to the axis of rotation. A_0, A_2 are on the same 'level' above this plane and the same applies to the three points B_0, B_2, C. Then $a_0 b_0 = a_2 b_2 = a_0 c$, showing that a rotation about an axis through A_0 parallel to the original axis would carry B_0 to C. (This proof does not apply to a diad rotatory inversion, which is, however, a simple reflexion and is covered by (iii) above.)

We now consider rotations. If a lattice has rotation of angle $2\pi/n$ about any axis it consists of nets perpendicular to that axis having the same rotation: for if lattice points A_0, B_0 are carried to A_1, B_1, the translations $A_0 A_1$ and $B_0 B_1$

Fig. 23.03

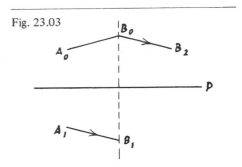

Fig. 23.04

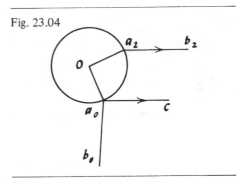

Fig. 23.02

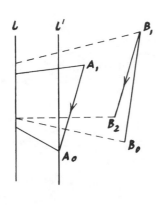

(a)

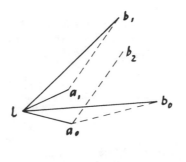

(b)

are perpendicular to the axis and determine, with A_0 as origin, a net of lattice points perpendicular to the axis. Moreover all lattice points in that plane form a net having the rotation $2\pi/n$; and since A_0 is any lattice point the whole lattice consists of such nets. As the rotations leave the nets unchanged, an axis of diad rotation must intersect the plane of a net either at a lattice point or at the mid-point of a line joining two of them. Combination with the translations then ensures that there are axes through all the lattice points. For triad rotations similar considerations apply.

Rotations of a net, and hence of a lattice, are restricted to the 'crystallographic' numbers 2, 3, 4 and 6; and if there are rotation axes in more than one direction, those through any one lattice point belong to a point group, the set of rotation numbers being one of:

$$2, 2, 2; \quad 3, 2, 2; \quad 4, 2, 2; \quad 6, 2, 2; \quad 3, 3, 2; \quad 4, 3, 2.$$

We choose as basic net one that allows the smallest rotation. For example, we shall consider the 4, 3, 2 possibility with a square net and the 6, 2, 2 with a hexagonal one.

We now consider the five types of net in turn.

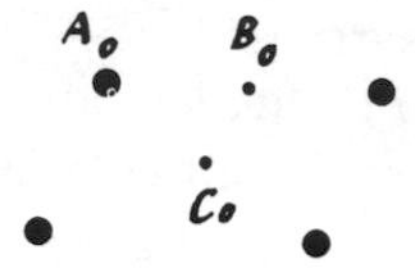

The parallelogram net

The general translation, applied to this net, produces a lattice whose only symmetry is inversion. This is the *triclinic* lattice, the three angles between edges of the unit cell being all of them oblique. But the parallelogram also has diad rotation, about such points as A_0, B_0 and C_0 in Fig. 23.05, that is to say about axes through those points perpendicular to the plane of the net. For the whole lattice to have this rotation the net in the nearest parallel plane must have its rotation axes in these same positions. Thus the point A_0 can be moved by the third translation either perpendicular to the net, reaching a point A_1 which we describe as 'opposite' to A_0, or to points opposite B_0 or C_0. The lattices so produced are shown in Fig. 23.06, the dots representing the points of a net and the rings those of the nearest parallel net. In (a) the unit cell has two pairs of rectangular faces; in (b) the same, but one pair are centred. (See also Fig. 23.14.) The lattice in (c) is essentially the same as that in (b), using a different parallelogram as base of the unit cell. These are the *monoclinic* lattices, only one of the three angles between edges of the unit cell being oblique. The lattice in (a) is primitive, that in (b) or (c) is centred. In both types there is reflexion in planes parallel to the original net, so the symbols are $P2/m$ and $B2/m$ (B being used rather than C, to indicate that the centring is on one of the rectangular faces of the unit cell, not on the basic parallelogram face).

Fig. 23.05

Fig. 23.06

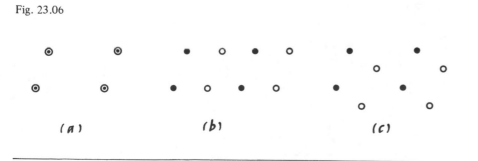

(a)   (b)   (c)

Fig. 23.07

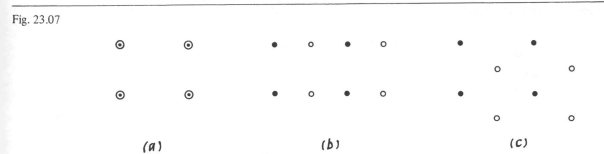

(a)   (b)   (c)

We cannot have diad axes in more than one direction with this net, because we know that in the point group whose rotation numbers are 2, 2, 2 the axes are in three mutually perpendicular directions.

The rectangular net

We can now have diad rotation about axes in three directions. Neighbouring nets can be related in the same three ways, as shown in Fig. 23.07. The (b) type gives a centred rectangular net and is more conveniently dealt with below. This leaves (a), which is a primitive lattice, and (c), which is body-centred. (See also Fig. 23.14.) Both types have reflexion in planes perpendicular to the three rotation axes. The full symbols are $P\frac{2}{m}\frac{2}{m}\frac{2}{m}$ and $I\frac{2}{m}\frac{2}{m}\frac{2}{m}$, where I stands for 'body-centred'. These are often abbreviated to $Pmmm$ and $Immm$.

The centred rectangular net

Here the (a) and (b) types are the only ones, the cell being thus centred on one pair of faces or all three pairs (Fig. 23.08). The symbols for these lattices are $C\frac{2}{m}\frac{2}{m}\frac{2}{m}$ and $F\frac{2}{m}\frac{2}{m}\frac{2}{m}$, where F stands for 'centred on all faces'. They are abbreviated to $Cmmm$ and $Fmmm$. These four rectangular lattices are grouped together under the heading *the orthorhombic system*.

The hexagonal net

This net allows hexad or triad rotation. Hexad axes must pass through the lattice points, but triads may equally well be through the centres of the triangles forming the net. So, for the lattice to have hexad rotation, there is no question about the direction of the third translation: it must be perpendicular to the plane of the net, with A_1 opposite A_0 (Fig. 23.09(a)). For triad rotation A_1 may be opposite the centre of one of the equilateral triangles, as in Fig. 23.09(b). But in that case a repetition of the movement brings A_1 to A_2, still

Fig. 23.08

Fig. 23.09

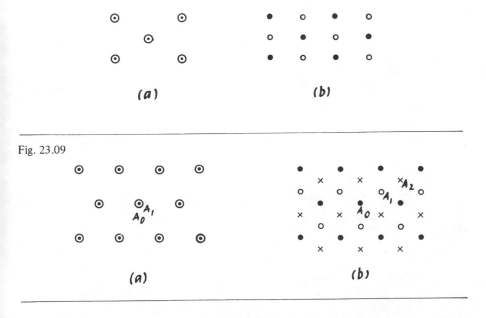

(a)   (b)

not opposite any point of the original net. The translation, which may be in any of three directions, has brought the original net, marked by dots, to the position shown by the rings, and thence to that of the crosses. A further repetition would bring it to a position opposite the original one. Thus the lattice might be described as built up from that of Fig. 23.09(*a*) by inserting two additional layers between each pair of neighbouring layers, the points of these intermediate layers being placed opposite points of trisection of the long diagonals of the rhombuses. Thus the lattice of Fig. 23.09(*b*) is related to that of Fig. 23.09(*a*) in somewhat the same way as a body-centred rectangular lattice to the primitive rectangular one from which it is derived.

The lattice of Fig. 23.09(*a*) is called *hexagonal* and its symbol is $P\frac{6}{m}\frac{2}{m}\frac{2}{m}$, often shortened to *P6/mmm*; while that of Fig. 23.09(*b*) is called *rhombohedral*, because it can be built up of cells with rhomboidal faces. Fig. 23.10 shows three of the faces of one cell. (See also Fig. 8.11(*c*), p. 48.) The symbol for this lattice is $R\bar{3}m$, because there is rotatory inversion about the triad axes and reflexion in planes containing them. The rotatory inversion is shown more clearly in Fig. 23.11, which is the same as Fig. 23.09(*b*), but with numerals to show the three parallel planes. Triad rotatory inversion in a point marked '1' about an axis perpendicular to the '1' plane would effect the sequence 202020 shown in the figure.

The symmetries of these two hexagonal lattices include point groups with the rotation numbers 6, 2, 2 and 3, 2, 2. We must also consider 3, 3, 2, corresponding to the rotation group of the regular tetrahedron. It might be thought that this would be a special case of the rhombohedral lattice, with the parallel planes so spaced that each lattice point would form a regular tetrahedron with three in the layer below. This, however, is not so, because the gaps between the tetrahedra would not conform to the desired symmetry. In fact they would be triangular anti-prisms, having symmetry $\bar{3}\frac{2}{m}$ (Fig. 5.12, p. 30). There is no lattice with the rotation numbers 3, 3, 2.

The square net
The net has tetrad rotation and the rotation numbers may be 4, 2, 2 or 4, 3, 2. If A_0 is one vertex of a square, a point A_1 in the nearest parallel net may be either opposite A_0 or opposite the centre of a square (Fig. 23.12(*a*), (*b*)). It cannot be opposite the mid-point of a side without destroying the tetrad symmetry. With general separation of the planes, the rotation numbers are 4, 2, 2 and the lattices are called *tetragonal*. The symbols are $P\frac{4}{m}\frac{2}{m}\frac{2}{m}$ and $I\frac{4}{m}\frac{2}{m}\frac{2}{m}$, often shortened to *P4/mmm* and *I4/mmm*.

Fig. 23.10

Fig. 23.11

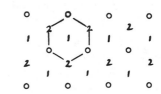

Fig. 23.12

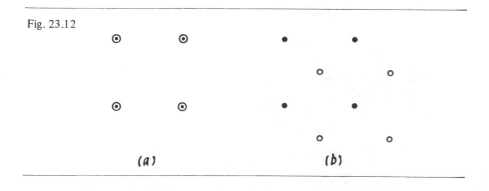

(*a*)   (*b*)

Fig. 23.13

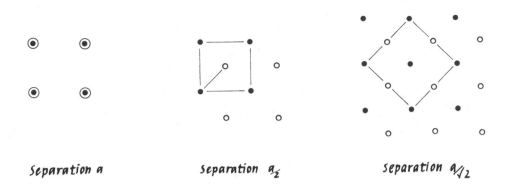

Separation a separation $a_{\overline{2}}$ separation $a_{\sqrt{2}}$

To obtain a cubic lattice, with rotation numbers 4, 3, 2, the planes must be separated at particular intervals, chosen so that there are further square nets perpendicular to the original ones. If the side of the square is a, the separation, in Fig. 23.12, could be either a or $a\sqrt{2}$. If it is a, the lattice is a primitive cubic one (Fig. 23.13(a)). It has the symmetries of the cube, which include triad rotatory inversion, and the symbol is therefore $P\frac{4}{m}\bar{3}\frac{2}{m}$, usually abbreviated to $Pm3m$. If the separation is $a\sqrt{2}$, the unit cell, though still cubic, is centred on one pair of faces. This destroys the triad symmetry and the lattice is of the tetragonal type $P4/mmm$.

In Fig. 23.12(b) a cubic lattice can be produced by a separation of $a/2$, giving a body-centred cube (Fig. 23.13(b)), or $a/\sqrt{2}$, giving a cube centred on all faces (Fig. 23.13(c). These three lattices all have triad rotatory inversion and their symbols are $P\frac{4}{m}\bar{3}\frac{2}{m}$, $I\frac{4}{m}\bar{3}\frac{2}{m}$ and $F\frac{4}{m}\bar{3}\frac{2}{m}$, usually abbreviated to $Pm3m$, $Im3m$ and $Fm3m$.

There is no need to consider the possibility of triad axes in other directions, because we proved (p. 126) that with the 4, 3, 2 rotation numbers the angles between the axes are as in the cube.

There are thus 14 lattices and no more. They are listed in Table 23.1 with full symbols, the shortened forms being given in parentheses. They are shown diagrammatically in Fig. 23.14 and, in more pictorial form, in the diagrams of Chapter 8.

Table 23.1

Triclinic	$P\bar{1}$			
Monoclinic	$P2/m$	$B2/m$		
Orthorhombic	$P\frac{2}{m}\frac{2}{m}\frac{2}{m}$	$C\frac{2}{m}\frac{2}{m}\frac{2}{m}$	$I\frac{2}{m}\frac{2}{m}\frac{2}{m}$	$F\frac{2}{m}\frac{2}{m}\frac{2}{m}$
	$(Pmmm)$	$(Cmmm)$	$(Immm)$	$(Fmmm)$
Tetragonal	$P\frac{4}{m}\frac{2}{m}\frac{2}{m}$	$I\frac{4}{m}\frac{2}{m}\frac{2}{m}$		
	$(P4/mmm)$	$(I4/mmm)$		
Hexagonal	$P\frac{6}{m}\frac{2}{m}\frac{2}{m}$			
	$(P6/mmm)$			
Rhombohedral	$R\bar{3}\frac{2}{m}$			
	$(R\bar{3}m)$			
Cubic	$P\frac{4}{m}\bar{3}\frac{2}{m}$	$I\frac{4}{m}\bar{3}\frac{2}{m}$	$F\frac{4}{m}\bar{3}\frac{2}{m}$	
	$(Pm3m)$	$(Im3m)$	$(Fm3m)$	

Fig. 23.14   The 14 Bravais lattices

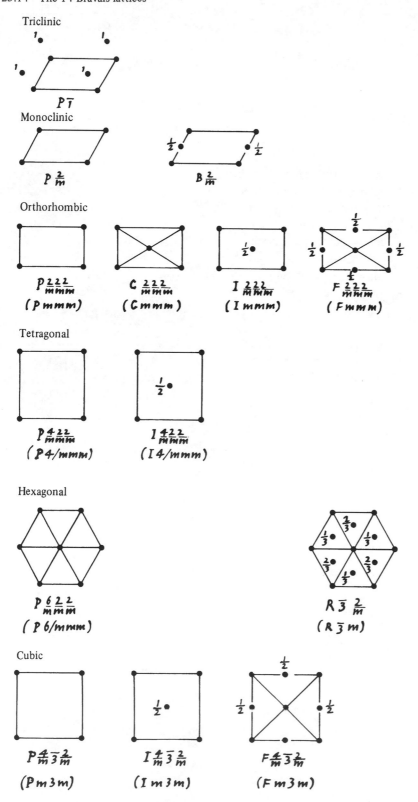

24

Space groups I

A *space group* is a group of movements in three dimensions that includes translations in three non-coplanar directions. It may also include rotations and screws, reflexions and glides, inversions and rotatory inversions. In this chapter and the next we enumerate the 230 space groups, deriving them from the 32 point groups. It is a long process, requiring care and patience rather than high mathematical skill, and not all readers will wish to follow it through in detail; but it is given here for completeness and for reference. We shall assume that we can obtain all the space groups by the methods used for line groups and plane groups, i.e. by combining the point groups with the appropriate lattices, sometimes replacing reflexions by glides or rotations by screws.

The lattices represent the translations of the space groups. They were divided into six systems according to the rotations they could accommodate, each system containing a primitive lattice and usually one or more derived lattices obtained by centring the unit cell. The hexagonal system does double duty, allowing both triad and hexad rotation, but its derived lattice, the rhombohedral, has triad only. It is therefore convenient to divide the system into two, calling it *trigonal* when only triad rotation is concerned, *hexagonal* when hexad rotation is relevant. This makes seven systems in all.

Each point group is associated with one of the seven systems, the one in fact whose lattices have all the symmetries of the point group and as few additional symmetries as possible. Table 24.1 gives the order of presentation.

Table 24.1

System	Symmetry of primitive lattice	Point groups						
Triclinic	$\bar{1}$	1	$\bar{1}$					
Monoclinic	$\dfrac{2}{m}$	2	m	$\dfrac{2}{m}$				
Orthorhombic	$\dfrac{2}{m}\dfrac{2}{m}\dfrac{2}{m}$	222	2mm	$\dfrac{2}{m}\dfrac{2}{m}\dfrac{2}{m}$				
Tetragonal	$\dfrac{4}{m}\dfrac{2}{m}\dfrac{2}{m}$	4	$\bar{4}$	$\dfrac{4}{m}$	422	4mm	$\bar{4}2m$	$\dfrac{4}{m}\dfrac{2}{m}\dfrac{2}{m}$
Trigonal (incl. rhombohedral)	$\dfrac{6}{m}\dfrac{2}{m}\dfrac{2}{m}$	3	$\bar{3}$	32	3m	$\bar{3}\dfrac{2}{m}$		
Hexagonal	$\dfrac{6}{m}\dfrac{2}{m}\dfrac{2}{m}$	6	$\bar{6}$	622	$\dfrac{6}{m}$	$\bar{6}2m$	6mm	$\dfrac{6}{m}\dfrac{2}{m}\dfrac{2}{m}$
Cubic	$\dfrac{4}{m}\bar{3}\dfrac{2}{m}$	23	$\dfrac{2}{m}\bar{3}$	432	$\bar{4}3m$	$\dfrac{4}{m}\bar{3}\dfrac{2}{m}$		

Reflexions or glides in two intersecting planes determine a rotation or screw about the line of intersection, or about a parallel line, as axis. It is therefore convenient, in dealing with groups in which reflexions or glides occur, to consider those movements first, allowing the rotations to appear as a consequence. For the diagrams the reflexion or glide planes, if any, will be taken as faces of the unit cells. The rotation axes may or may not pass through the origin.

Coordinate axes will be taken along the edges of the unit cell, with the lengths of the edges as units for the coordinates. In the diagrams the y-axis will be drawn 'horizontally' across the page, with the x-axis down the page. The z-axis is then perpendicular to the page (except in the triclinic system, when it is at an oblique angle). It is customary for the z-axis to be taken as the unique axis, if there is one. In the symbols for the symmetries of the groups it is convenient to take the z-axis first, but the practice is by no means universal. (In the *International Tables for X-ray Crystallography* the order x, y, z is used for the orthorhombic system, though not for the others. We prefer to put the z-axis first in all cases, especially as even in the orthorhombic system there is sometimes a unique axis.)

The planes through the origin perpendicular to the z-, x- and y-axes will be called the Z-, X- and Y-planes respectively. The first position in each symbol (after the prefix indicating whether the lattice is primitive or centred) will always refer to the z-axis and the Z-plane.

Centred lattices will be given the prefixes C, A and B, when they are centred in the xy-, yz- or zx-planes. If the unit cell is centred on all faces the prefix will be F and if it is body-centred it will be I. Glides in the direction of the axes will be indicated by c, a and b respectively, and n will indicate a diagonal glide. A 'diamond' glide (to be explained later) will be indicated by d.

Each diagram will show a unit cell, with a set of points arranged to indicate the symmetries. It is to be understood that the cell and the points are repeated at unit intervals in the directions of the three axes. The coordinate x will in fact represent $n + x$, where n is any integer. For this reason we shall always assume that z, x and y lie between 0 and 1, and that $-x$ is equivalent to $1 - x, \frac{1}{2} + x$ to $-\frac{1}{2} + x$, and so on.

If z is the z-coordinate of a point marked by a dot, a ring is used to represent the point with coordinate $-z$. If $\frac{1}{2}$ or other fraction is placed alongside a dot or ring, that fraction must be added to the z-coordinate. Thus a ring with $\frac{1}{2}$ alongside means a point with coordinate $\frac{1}{2} - z$. Dots and rings in red are reflected versions of the points in black.

The appearance of any of these diagrams depends very much on the choice of origin. When possible, we take the origin (shown at the top left-hand corner of the diagram) at the intersection of three reflexion or glide reflexion planes, or on the line of intersection of two such planes.

We now consider the 32 point groups in turn, arranged according to the seven systems.

The triclinic system
Point groups 1 and $\bar{1}$

There is only one type of lattice and, with no rotations or reflexions, each of these point groups gives just one space group (Fig. 24.01).

Fig. 24.01

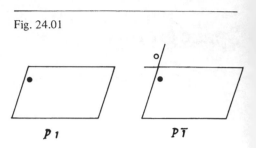

P 1 $P \bar{1}$

The monoclinic system

Point groups 2, m and 2/m

The z-axis is unique and there are two lattices $P2/m$ and $B2/m$. The diad rotation can be replaced by a 2_1 screw and the reflexion in the Z-plane by a glide. With the primitive lattice all directions in the Z-plane are on a par; and with the centred lattice the only distinguishable one is that of the centred face. The space groups are shown in Fig. 24.02.

The orthorhombic system

Point group 222

Here we have movements that react on one another. As mentioned before, diad rotations about two axes at right angles imply a third diad axis perpendicular to both. If the first two intersect, the third passes through the same point. This can be seen either from a diagram (Fig. 24.03) or by algebra, as follows:

coordinates of any point	z,	x,	y;
rotate about z-axis	z,	$-x$,	$-y$;
rotate about x-axis	$-z$,	$-x$,	y.

But $(-z, -x, y)$ is (z, x, y) rotated about the y-axis.

Fig. 24.03

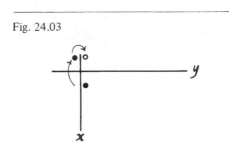

Fig. 24.02

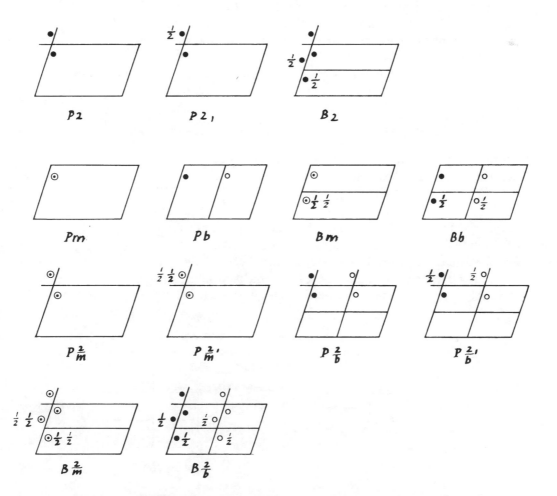

If the axes of rotation do not intersect, it appears, by either method, that the lattice must be body-centred; see Fig. 24.04, or as follows:

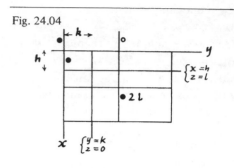

Fig. 24.04

coordinates of any point	z,	x,	y;
rotate about z-axis	z,	$-x$,	$-y$;
rotate about $y = k, z = 0$	$-z$,	$-x$,	$2k + y$;
rotate about $x = h, z = l$	$2l + z$,	$2h + x$,	$2k + y$.

With a primitive lattice there would also be a point at $(z, x, 2k + y)$ and hence a rotation about the y-axis, contrary to hypothesis; so it must be a body-centred lattice with unit cell of edges $4l$, $4h$, $4k$. The displacements l, h, k are therefore $\frac{1}{4}$ unit each.

If one of two rotations about intersecting axes is replaced by a screw, the axis of the resultant rotation is displaced $\frac{1}{4}$ unit in the direction of the screw. Thus, for example:

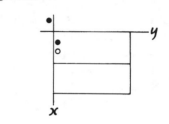

Fig. 24.05

coordinates of any point	z,	x,	y;
rotate about z-axis	z,	$-x$,	$-y$;
screw about x-axis	$-z$.	$\frac{1}{2} - x$,	y;

As also shown in Fig. 24.05, this represents a rotation about the line $z = 0$, $x = \frac{1}{4}$.

If the unit cell is centred on one face, rotation axes lying in that face imply screw rotations about intermediate parallel axes (Fig. 24.06(a)) and vice versa (Fig. 24.06(b)). The symbol 2 is used in such cases. Screw axes without parallel rotation axes are designated 2_1 and can only occur at right angles to the centred face. This limits the number of space groups containing 'independent' screw rotations (the 2_1 type), and in fact prohibits them entirely with the F-centred and, for similar reasons, the I-centred lattices.

We can now enumerate the space groups based on the point group 222. With the primitive lattice there is no distinction between the three axes; and we are able to have all four combinations of rotations and screws, namely:

$$P222, \quad P2_1 22, \quad P22_1 2_1, \quad P2_1 2_1 2_1.^*$$

With the body-centred lattice the rotation axes can be intersecting or non-intersecting. These space groups are shown in Fig. 24.07.

Point group 2mm

It will be recalled that reflexions in two planes intersecting at right angles combine to give a diad rotation about the line of intersection; and that if one reflexion is replaced by a glide the effect, if the glide is in the direction of the axis, is to change the rotation to a screw. If, however, it is in the perpendicular direction it displaces the axis $\frac{1}{4}$ unit in that direction. It is helpful here to think

*We continue to take the axes in the order z, x, y.

Fig. 24.06

(a)   (b)

in terms of translations added to the rotation. With reflexions in the X- and Y-planes the rotation is about the z-axis. A glide in either plane in the z-direction (a c-glide) adds half a unit in that direction, changing 2 to 2_1; or if there are c-glides in both planes a whole unit is added, which has no effect. Glides in the perpendicular directions (b in the X-plane or a in the Y-plane) displace the rotation axis $\frac{1}{4}$ unit in their own directions. A diagonal glide (n) has both effects.

With a primitive lattice the possibilities are therefore as shown in Table 24.2. The omission of six possibilities in the lower left-hand part of this table is due to the fact that the x- and y-axes are indistinguishable. These ten space groups are illustrated in Fig. 24.08.

With centred lattices the number of possibilities is reduced by *pairing*. To explain this we note first that in all the above space groups the reflexions or glides in intermediate planes are the same as in the parallel principal planes. But with centred lattices this is no longer true. If the lattice is centred on one face of the unit cell (C-centred, for example) a reflexion in one of the other planes (which we shall call the 'side planes') implies glide reflexions in intermediate parallel planes (Fig. 24.09(a)); and vice versa, a glide implies reflexions

Table 24.2

$P2mm$	$P2_1mc$	$P2ma$	$P2_1mn$
	$P2cc$	$P2_1ca$	$P2cn$
		$P2ba$	$P2_1bn$
			$P2nn$

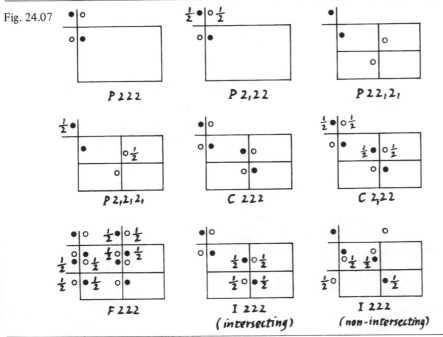

Fig. 24.07

$P222$ $P2_122$ $P22_12_1$

$P2_12_12_1$ $C222$ $C2_122$

$F222$ $I222$ (*intersecting*) $I222$ (*non-intersecting*)

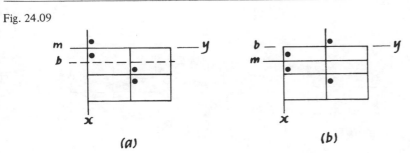

Fig. 24.09

(a)   (b)

Fig. 24.08

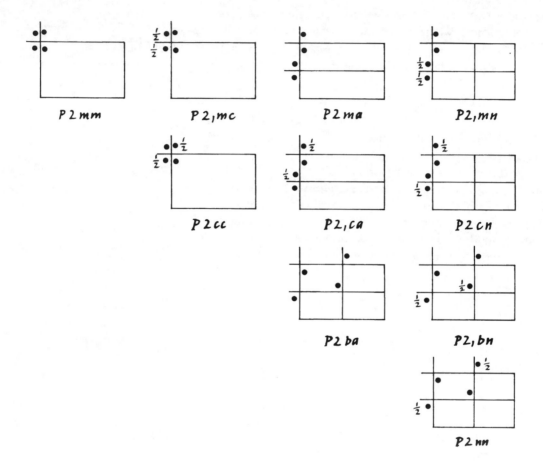

P2mm   P2₁mc   P2ma   P2₁mn

P2cc   P2₁ca   P2cn

P2ba   P2₁bn

P2nn

(Fig. 24.09(*b*)). So with the *C*-centred lattice, *m* and *b* are alternative descriptions. We say that these letters are 'paired'.

In a similar way a glide (in one of the side planes) in a direction perpendicular to the centred plane implies a diagonal glide in intermediate planes (Fig. 24.10(*a*)) and vice versa (Fig. 24.10(*b*)). The letters *c* and *n* are 'paired', like *m* and *b*.

The space group represented in Fig. 24.11 could be described in any of four ways:

$$C2_1mc, \quad C2_1mn, \quad C2_1bc, \quad C2_1bn.$$

Fig. 24.10

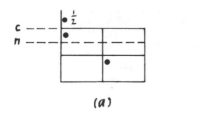

(a)

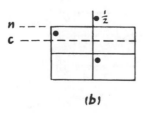

(b)

Fig. 24.11

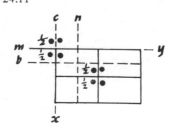

We need only take one letter from each pair. So, for a *C*-centred lattice, we need only use *m* and *c* in the side planes and, as the side planes are indistinguishable, there are only three types, namely 2*mm*, 2*mc* and 2*cc* (Fig. 24.12).

For glides in the centred plane, those in the two principal directions form a 'pair' (one implies the other) and a diagonal glide pairs with a simple reflexion. This is shown in Fig. 24.13(*a*) for an *A*-centred lattice and in Fig. 24.13(*b*) for a *C*-centred one. (This latter will occur with the point group $\frac{2}{m}\frac{2}{m}\frac{2}{m}$.)

So for an *A*-centred lattice we need only use *m* and *b* in the *X*-plane, *m* and *a* in the *Y*-plane. There are thus four space groups, as shown in Fig. 24.14.

Fig. 24.12

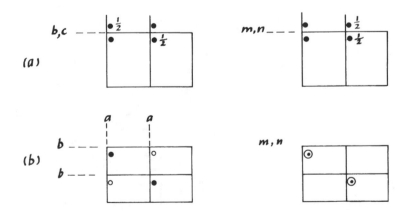

C 2mm   C₁2mc   C 2cc

Fig. 24.13

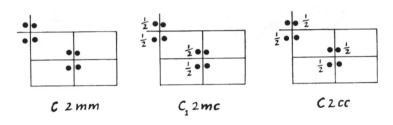

(a)

(b)

Fig. 24.14

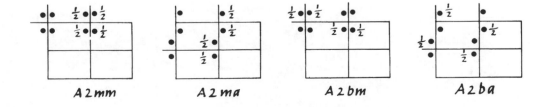

A 2mm   A 2ma   A 2bm   A 2ba

With an *I*-centred (body-centred) lattice the rule is the same as for glides in the plane of a centred face. Thus in the *X*-plane the pairs are *m* and *n*, *c* and *b*; and in the *Y*-plane *m* and *n*, *c* and *a*. There is again no distinction between the *X*- and *Y*-planes, so there are only three space groups, as shown in Fig. 24.15.

With an *F*-lattice (centred on all faces) any one of these glides is parallel to one face and perpendicular to each of the others. The consequence is that a reflexion *m* in any plane pairs off in turn with each of the three glides (e.g. in the *X*-plane with *b*, *c* and *n*). So far as these glides are concerned there is only one space group (Fig. 24.16).

But there is another kind of glide, not hitherto considered, called a *diamond glide*. This is a glide along the diagonal of a centred face or diagonal plane of a cell. Any glide, when repeated, gives a translation, and usually this has been from one corner of a unit cell to another corner. But if the cell is centred it may be from one corner to a face-centre, or to the body-centre. The glide distance must then be a quarter of the diagonal length. Fig. 24.17(*a*) shows such a glide in the *X*-face of a unit cell, the black dot marked $\frac{1}{2}$ being near the centre of that face. This presupposes that the point group from which the space group was constructed contained reflexion in that face. (That is why diamond glides do not occur in the monoclinic system.) With the 2*mm* group a face-

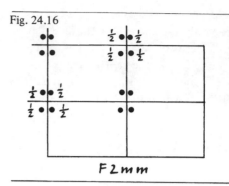

Fig. 24.16

F 2 m m

Fig. 24.15

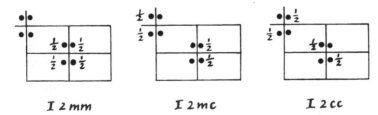

I 2 mm   I 2 mc   I 2 cc

Fig. 24.17

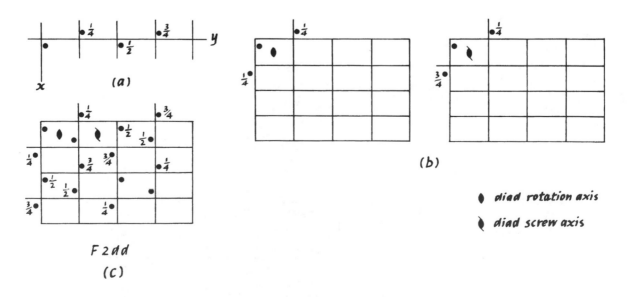

(a)

(b)

F 2 dd

(c)

● *diad rotation axis*

◗ *diad screw axis*

centre can be used in this way, but not a body-centre, because there is no reflexion in a diagonal plane containing the body-centre. (With the 4*mm* group we shall find it possible to use the body-centre.) With the 2*mm* group the lattice must be *F*-centred, because a diad rotation or screw about the z-axis combines with the glide to make a similar glide in the *Y*-plane (Fig. 24.17(*b*)). As will be seen from the figure, the rotation or screw axis is displaced $\frac{1}{8}$ unit in both directions from the intersection of the glide planes and the completion of the figure gives the same result in both cases (Fig. 24.17(*c*)), with parallel glide planes at quarter-unit intervals, the glides being alternately in different directions.

Point group $\frac{2}{m}\frac{2}{m}\frac{2}{m}$

The space groups can be found in the same way. In each of the three principal planes there is a reflexion which may be replaced by a glide in any of three directions (two principal and one diagonal). But there is no unique axis and this considerably reduces the number of possibilities. With a primitive lattice there are in fact 16, as shown in Table 24.3. The rotations again depend on the reflexions and follow the same rules as before. In the z-direction, for example, the rotation becomes a screw if one, but not both, of the letters for the *X*- and *Y*-planes is *c* or *n*; and the rotation axis is displaced in one or both directions if one or both of those letters is *b*, *a* or *n*. These groups are illustrated in Fig. 24.18.

With a *C*-centred lattice the letters are paired as follows:

in the *Z*-plane   *m, n*; *a, b*   (see Fig. 24.13(*b*)):
in the *X*-plane   *m, b*; *c, n* ⎫
in the *Y*-plane   *m, a*; *c, n* ⎭ (see Figs 24.19 and 24.20).

Picking the first letter of each pair and bearing in mind that the x- and y-axes are indistinguishable, we have the following six combinations:

mmm, mmc, mcc, amm, amc, acc.

The rotations and screws depend as before on these reflexions and glides, but a screw about an axis in a centred plane is necessarily accompanied by a rotation about a parallel axis (see Fig. 24.06), so in such cases 2_1 becomes 2, the rotation axis being displaced, as shown in Fig. 24.19.

With the *I*-centred lattice the pairings are different:

in the *Z*-plane   *m, n*; *a, b*;
in the *X*-plane   *m, n*; *b, c*;
in the *Y*-plane   *m, n*; *c, a*.

All possibilities are covered by the four combinations:

mmm, amm, mcc, acc.

The space groups are shown in Fig. 24.20.

Table 24.3

mmm	*nmm*	*mnn*	*nnn*
	amm	*mna*	*ann*
		mnc	*ncc*
		mcc	*nca*
		mca	*nba*
		mba	*acc*
			abc

Fig. 24.18

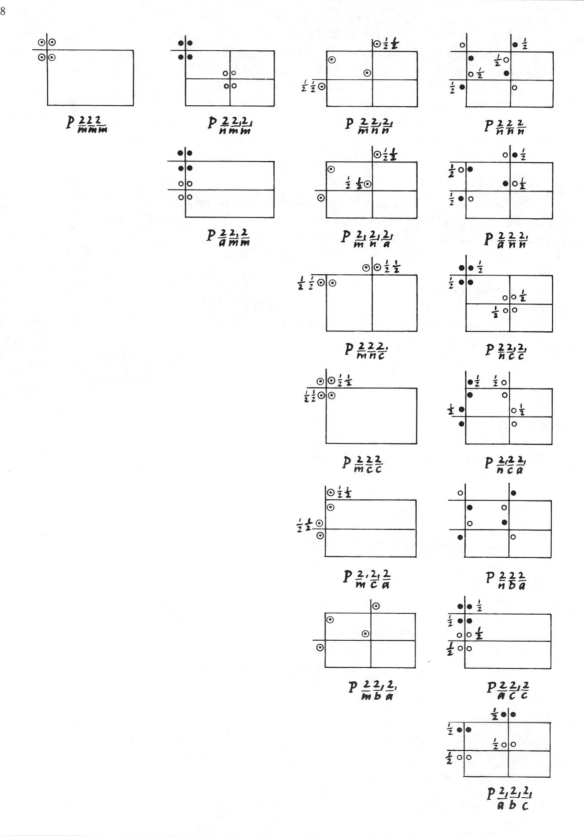

With the *F*-centred lattice there are again only two groups, one with reflexions (Fig. 24.21(*a*)) and one with diamond glides. This latter is shown in Fig. 24.21(*b*) with the origin at the intersection of three glide planes, the rotation axes intersecting at $(\frac{1}{8}, \frac{1}{8}, \frac{1}{8})$ and similar points. But a rather neater representation is obtained by taking the origin at the intersection of three rotation axes, as in Fig. 24.21(*c*).

Fig. 24.19

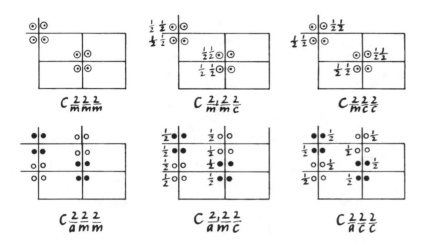

$$C\frac{2}{m}\frac{2}{m}\frac{2}{m} \qquad C\frac{2_1}{m}\frac{2}{m}\frac{2}{c} \qquad C\frac{2}{m}\frac{2}{c}\frac{2}{c}$$

$$C\frac{2}{a}\frac{2}{m}\frac{2}{m} \qquad C\frac{2_1}{a}\frac{2}{m}\frac{2}{c} \qquad C\frac{2}{a}\frac{2}{c}\frac{2}{c}$$

Fig. 24.20

$$I\frac{2}{m}\frac{2}{m}\frac{2}{m} \qquad I\frac{2}{a}\frac{2}{m}\frac{2}{m} \qquad I\frac{2}{m}\frac{2}{c}\frac{2}{c} \qquad I\frac{2}{a}\frac{2}{c}\frac{2}{c}$$

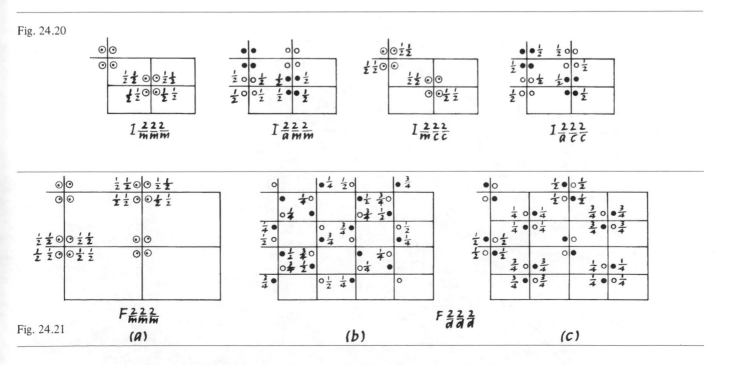

$$F\frac{2}{m}\frac{2}{m}\frac{2}{m} \qquad\qquad F\frac{2}{d}\frac{2}{d}\frac{2}{d}$$

Fig. 24.21   (*a*)   (*b*)   (*c*)

The tetragonal system

In this system there are only two kinds of lattice, P and I. (C pairs with P and F with I.) In the symbols, the first position after the prefix refers as before to the z-axis and the Z-plane, but the second position refers to both x- and y-axes and the corresponding planes, leaving the third position for the diagonal directions.

Point group 4

The tetrad rotation can be replaced by screws 4_1, 4_2 and 4_3, of which 4_1 and 4_3 are an enantiomorphic pair (left-handed and right-handed versions of the same pattern). With the I-lattice, 4_3 pairs with 4_1 and 4_2 with 4, so we consider only 4 and 4_1. There are thus six space groups, as shown in Fig. 24.22. (To see the pairings, note the screw axes, in the last two diagrams, through the points $x = 0, y = \frac{1}{2}$.)

Point group $\bar{4}$

There are only two groups, $P\bar{4}$ and $I\bar{4}$, as shown in Fig. 24.23. (Screws give nothing new, merely displacing the inversion centres.)

Fig. 24.22

Fig. 24.23

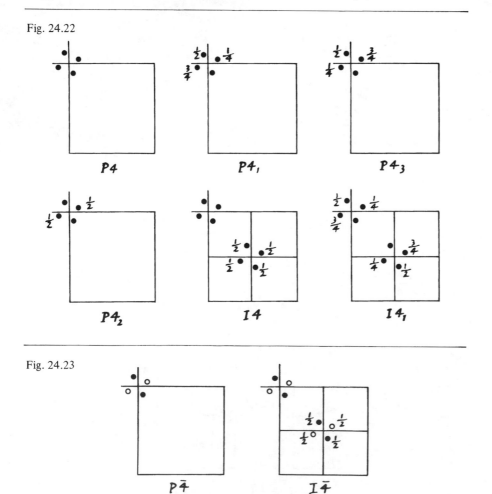

Point group 4/m

We first suppose that there is simple reflexion in the Z-plane. This excludes the screws 4_1 and 4_3, as they do not reflect. With a primitive lattice 4 and 4_2 can both be used (see Fig. 24.25, below), but with a body-centred lattice they pair. Thus we have $P4/m$, $P4_2/m$ and $I4/m$. If the reflexion is replaced by a glide, the tetrad rotation or screw ensures that there are glides in two perpendicular directions, either those of the x- and y-axes or those of the diagonals. But the first of these two possibilities leads to a centred lattice, which is merely a smaller lattice turned through 45° (Fig. 24.24). Thus the two possibilities are the same and we have only $P4/n$ and $P4_2/n$. (With an I-lattice the glides reduce to simple reflexions.) We must also consider a 4_1 screw with a glide. In this case the screw transforms the glide to one in a perpendicular direction in a parallel plane $\frac{1}{4}$ unit away. The combination of these two glides gives an I-centred lattice and the symbol for the group is $I4_1/g$. ($I4_3/g$, $I4_1/n$ and $I4_3/n$ would be the same.) The six space groups we have found are shown in Fig. 24.25. The glide planes for the last diagram have been taken at $z = \frac{1}{8}$, $z = \frac{3}{8}$, etc.

Fig. 24.24

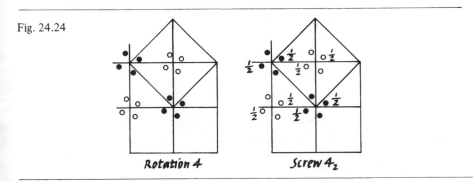

Rotation 4   Screw 4_2

Fig. 24.25

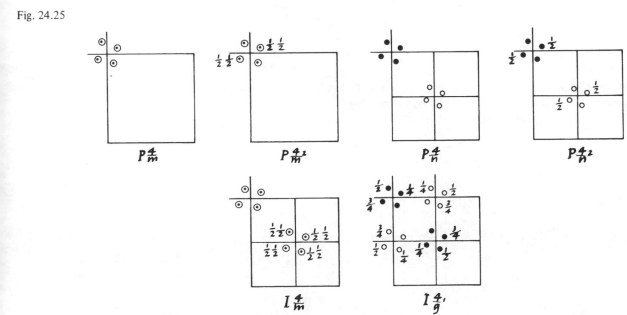

$P\frac{4}{m}$ $P\frac{4_2}{m}$ $P\frac{4}{n}$ $P\frac{4_2}{n}$

$I\frac{4}{m}$ $I\frac{4_1}{g}$

Fig. 24.26

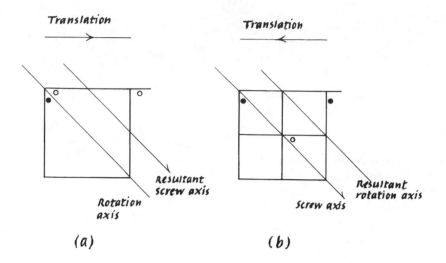

(a)   *(b)*

Point group 422

As already mentioned, the 4 in the first position refers to the z-axis, the 2 in the second position means diad rotation about both the x- and y-axes, and the 2 in the third position refers similarly to both the diagonal directions. The three numbers are interdependent, since diad rotations about axes at 45° to one another imply tetrad rotation about a perpendicular axis.

Diad rotation about a diagonal combines with translation along a side of the square to give a screw about an intermediate parallel axis (Fig. 24.26(a)) and vice versa (Fig. 24.26(b)). So, in the diagonal directions, 2 and 2_1 are paired. We use the symbol 2. In the other directions there are eight possible combinations of rotations and screws and, with a primitive lattice, these all give different space groups:

$$P422, \quad P4_1 22, \quad P4_2 22, \quad P4_3 22,$$
$$P42_1 2, \quad P4_1 2_1 2, \quad P4_2 2_1 2, \quad P4_3 2_1 2.$$

They are illustrated in Fig. 24.27. With screw rotations 4_1 or 4_3 the axes in the x- and y-directions do not intersect, because they are related by the screw rotation.

With a body-centred lattice, 4_2 pairs with 4 and 4_3 with 4_1. Moreover in the x- and y-directions 2_1 pairs with 2. Thus there are only two space groups, $I422$ and $I4_1 22$ (Fig. 24.28). The 4_2 axes in the first of these diagrams and the 4_3 axes in the second are such lines as $x = 0$, $y = \frac{1}{2}$ and $x = \frac{1}{2}$, $y = 0$.

Point group 4mm

With a primitive lattice the first of the two letters in the symbol can be m, c, g or n (where g refers to both x- and y-directions, n to both diagonals); but in the diagonal planes a horizontal glide pairs with m and an n-glide with c.

Fig. 24.28

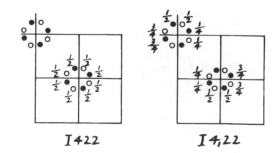

$I422$ $I4_1 22$

We therefore take the second letter as either *m* or *c*. This gives eight possible groups:

     mm,   *cm*,   *gm*,   *nm*,

     mc,   *cc*,   *gc*,   *nc*.

These reflexions or glides determine the tetrad movement, which is 4_2 if one (but not both) of the letters is *n* or *c*. The groups are shown in Fig. 24.29. The diagrams are drawn with the origin at the intersection of three reflexion or glide planes. This indeed is the easiest and most natural way to draw them, in view of the fact that the rotations are determined by the reflexions and glides. But

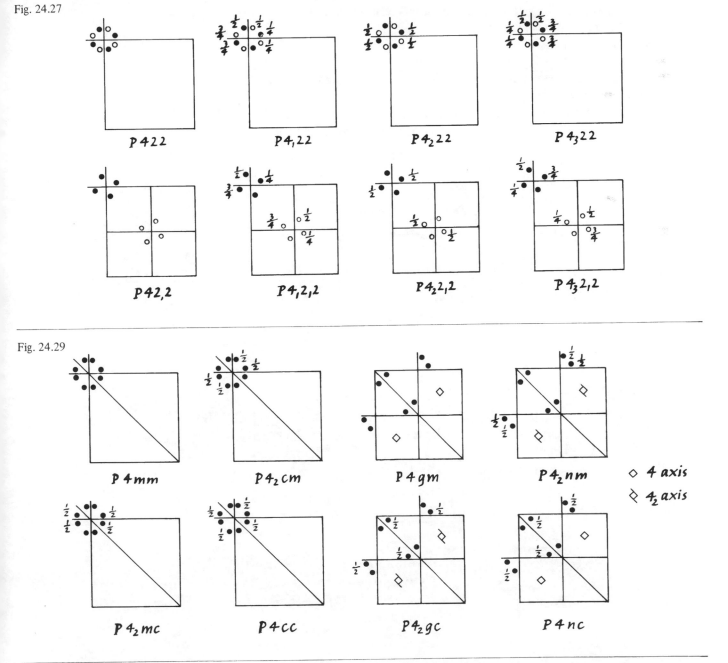

Fig. 24.27

$P422$ $P4_122$ $P4_222$ $P4_322$

$P42_12$ $P4_12_12$ $P4_22_12$ $P4_32_12$

Fig. 24.29

$P4mm$ $P4_2cm$ $P4gm$ $P4_2nm$

◇ 4 axis

⬦ 4_2 axis

$P4_2mc$ $P4cc$ $P4_2gc$ $P4nc$

some of them have a neater appearance if drawn with the origin on a tetrad axis, as in Fig. 24.30.

With a body-centred lattice the pairings in the X- and Y-, or parallel, planes are n with m, g with c; and in diagonal planes each of the three glides pairs with m. So, as far as these movements are concerned, there are only two groups, $4mm$ and $4cm$. It is, however, possible to have diamond glides in the diagonal planes and this gives two more. The four groups are shown in Fig. 24.31.

Point group $\bar{4}2m$ (including $\bar{4}m2$)

With the primitive lattice there are pairings in the diagonal planes, i.e. for the third position in the symbol. Here 2_1 pairs with 2, as before, and also the horizontal glide pairs with m, and n with c. This leaves eight possibilities for the second and third positions in the symbol:

$$2m, \quad 2c, \quad 2_1m, \quad 2_1c,$$
$$m2, \quad c2, \quad g2, \quad n2.$$

As shown in Fig. 24.32, these all lead to space groups, with $\bar{4}$ (tetrad rotatory inversion) following automatically in every case.

With the body-centred lattice there are pairings in planes in the x- and y-directions as well as diagonally. For the second position, 2_1 pairs with 2, n

Fig. 24.30

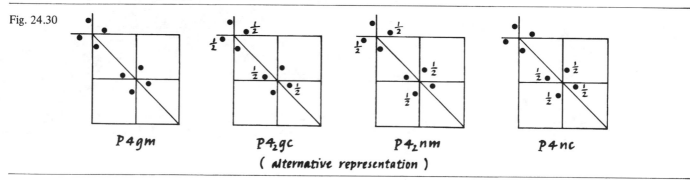

P4gm   P4₂gc   P4₂nm   P4nc
          (Alternative representation)

Fig. 24.31

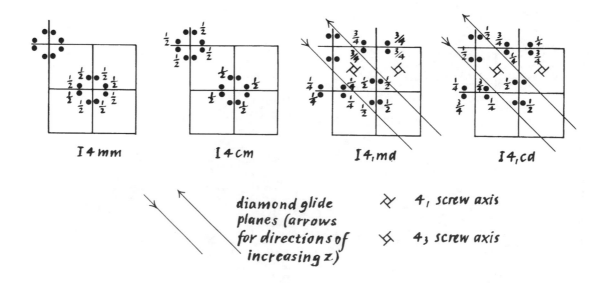

I4mm   I4cm   I4₁md   I4₁cd

diamond glide planes (arrows for directions of increasing z)

4₁ screw axis

4₃ screw axis

with m, and g with c; while in the third position c, g and n all pair with m. This leaves only $2m$, $m2$ and $c2$ (Fig. 24.33). But there is still the possibility of diamond glides. With the body-centred lattice these must be in planes perpendicular to the square bases of the cells and containing the diagonals of those bases. The directions of the glides must be those of the diagonals of the cells. With origin at a point where a diad axis, taken as x-axis, meets a glide plane, it

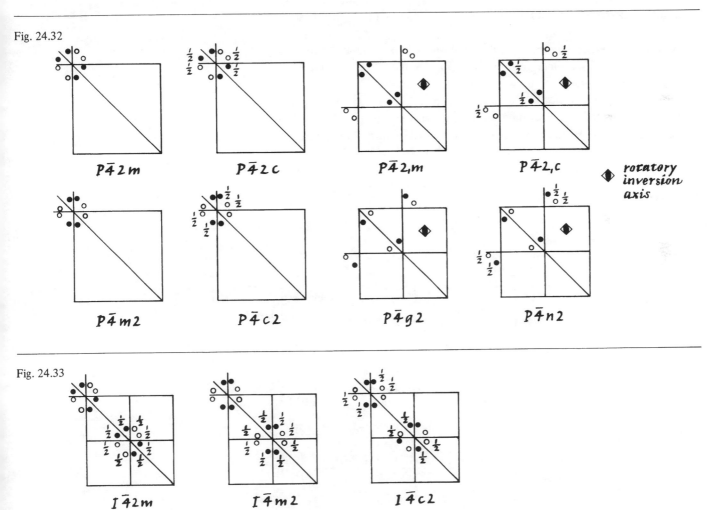

Fig. 24.32

$P\bar{4}2m$ $P\bar{4}2c$ $P\bar{4}2_1m$ $P\bar{4}2_1c$ ◆ rotatory inversion axis

$P\bar{4}m2$ $P\bar{4}c2$ $P\bar{4}g2$ $P\bar{4}n2$

Fig. 24.33

$I\bar{4}2m$ $I\bar{4}m2$ $I\bar{4}c2$

can be seen from Fig. 24.34(a) that the diad rotation transforms a glide in the direction $z = x = y$ to one in the direction $-z = x = -y$, i.e. $z = -x = y$. (In the figures the arrows indicate the directions in which z increases.) We then note that the glide transforms the diad axis $y = 0$, $z = 0$ to a non-intersecting diad $x = \frac{1}{4}$, $z = \frac{1}{4}$ (Fig. 24.34(b)). Continuing, we have the pattern of which one cell is shown in Fig. 24.34(c) or, with a different origin, in Fig. 24.34(d).

Point group $\frac{4}{m}\frac{2}{m}\frac{2}{m}$

The space groups can be easily derived from those of the point group $4mm$, by adding a reflexion or glide in the Z-plane. From each of the eight groups illustrated in Fig. 24.29 we can obtain two more by adding m or n in the first position:

mmm,   mcm,   mgm,   mnm,   nmm,   ncm,   ngm,   nnm,
mmc,   mcc,   mgc,   mnc,   nmc,   ncc,   ngc,   nnc.

These sixteen groups are shown in Fig. 24.35.

In these groups we do not have g in the first position because that would produce a centred lattice. With the body-centred lattice (the only centred one in this system) g is possible, provided the glides in the x-direction and in the y-direction are in planes $\frac{1}{4}$ unit apart (e.g. $z = \frac{1}{8}$ and $z = \frac{3}{8}$). This does in fact happen in two of the groups that we now consider.

Fig. 24.34

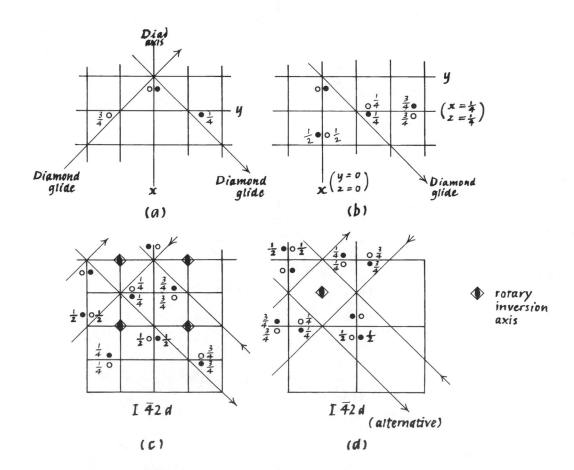

(a)   (b)

I $\bar{4}2d$ I $\bar{4}2d$
                        (alternative)

(c)   (d)

◆ rotary inversion axis

To obtain the body-centred groups we have to add a reflexion or glide, in the Z-plane or parallel to it, in each of the four groups shown in Fig. 24.31. In the first two, *n* (in the first position) pairs with *m*, and *g* is impossible because the diagonal reflexion would require the glides in the *x*- and *y*-directions to be in the same plane. In the other two, however, the diamond glides similarly

Fig. 24.35

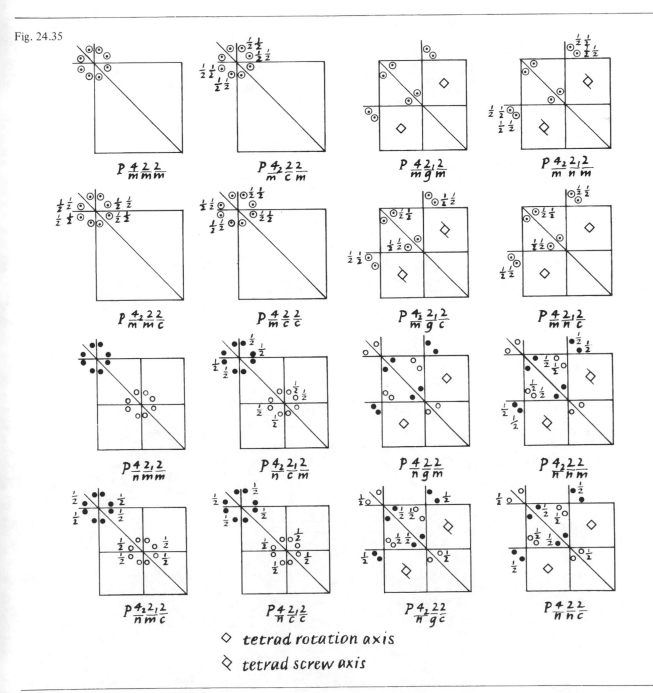

◇ tetrad rotation axis

◈ tetrad screw axis

exclude *m*, and *g* is the only possibility. There are therefore four groups, based on

$$mmm, \quad mcm, \quad gmd, \quad gcd,$$

as illustrated in Fig. 24.36. In the last two groups there are glides in the *x*-direction in the planes $z = \frac{1}{8}$, $z = \frac{5}{8}$, and in the *y*-direction in the planes $z = \frac{3}{8}$, $z = \frac{7}{8}$.

Fig. 24.36

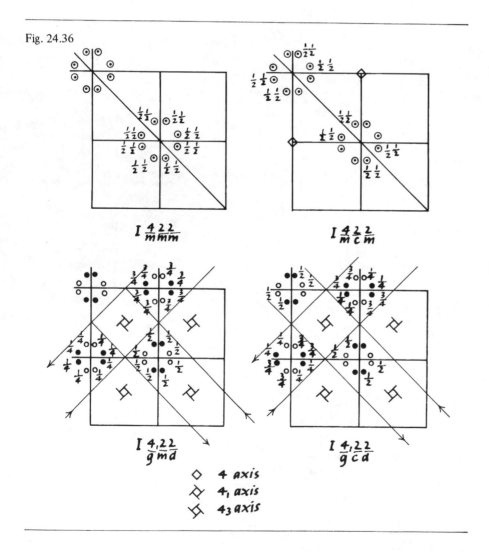

$$I\frac{4}{m}\frac{2}{m}\frac{2}{m} \qquad\qquad I\frac{4}{m}\frac{2}{c}\frac{2}{m}$$

$$I\frac{4_1 2 2}{g\,m\,d} \qquad\qquad I\frac{4_1 2 2}{g\,c\,d}$$

◇   **4** *axis*

⬦   **4₁** *axis*

⬦   **4₃** *axis*

25

Space groups II

We have dealt so far with the four systems in which the only rotations are diad and tetrad. We now go on to those in which triad or hexad rotations occur, namely the trigonal, hexagonal and cubic systems. Of these the trigonal and hexagonal have much in common. They are both based on the same primitive lattice, with symmetry $\frac{6}{m}\frac{2}{m}\frac{2}{m}$. But there is a derived lattice, the rhombohedral, which does not have the hexagonal symmetry or reflexions in the Z-plane and parallel planes. The distinction between the trigonal and hexagonal systems is that the trigonal includes only those point groups that can be applied also to the rhombohedral lattice. The points of that lattice are arranged as in Fig. 25.01, and its symmetries are $\bar{3}\frac{2}{m}$. So the point groups available are 3, $\bar{3}$, 32, 3m and $\bar{3}\frac{2}{m}$. (3/m does not belong, and that is why it is more suitably designated by its alternative symbol $\bar{6}$.)

The trigonal system
The only screws are about axes in the z-direction: for, as shown in Fig. 25.02(a), diad screws about axes in the Z-plane at 120° to one another would combine to give a translation of half a unit. Similarly the only glides are in the z-direction, for g-glides (Fig. 25.02(b)) or n-glides (Fig. 25.02(c)) in planes containing the z-axis would again lead to half-unit translations. (Similar arguments apply for glides in planes bisecting the angles between those considered.)

Fig. 25.01

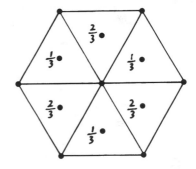

Fig. 25.02

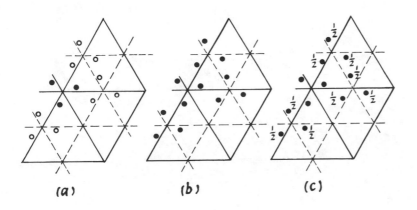

(a)   (b)   (c)

Point group 3

With a primitive lattice the triad rotation can be replaced by a screw, 3_1 or 3_2; but with the rhombohedral lattice either of these screws implies a simple rotation. There are thus four space groups

$$P3, \quad P3_1, \quad P3_2, \quad R3,$$

as shown in Fig. 25.03.

Point group $\bar{3}$

There are only two space groups, $P\bar{3}$ and $R\bar{3}$ (Fig. 25.04).

Point group 32

There are three diad axes, related by the triad rotation. It is convenient, in both the trigonal and hexagonal systems, to have three coordinate axes (x, y and u) in the Z-plane, related by the triad rotation. We take them, as in Fig. 25.05, along sides of the smallest triangles of the lattice. The diad axes may then be either parallel to these coordinate axes or parallel to the bisectors of the angles formed by them. The symbols for these types are 32 and 312 respectively, the number in the second position referring to the directions of the coordinate axes. With the primitive lattice, the triad rotation can be replaced by a screw,

Fig. 25.05

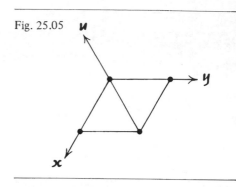

Fig. 25.03

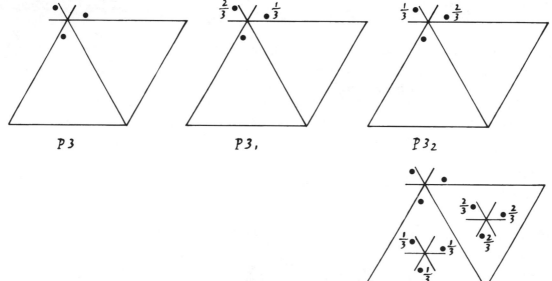

Fig. 25.04

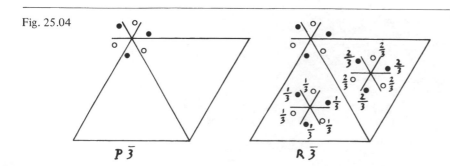

either 3_1 or 3_2, giving six space groups; with the rhombohedral lattice, screws pair with simple rotations and the diad axes can only be in the x, y and u directions. (It can be seen from Fig. 25.01 that the lattice itself does not allow rotation about the bisectors.) There are thus seven space groups altogether:

$P32$, $P3_12$, $P3_22$,
$P312$, $P3_112$, $P3_212$,
$R32$.

These are shown in Fig. 25.06.

Fig. 25.06

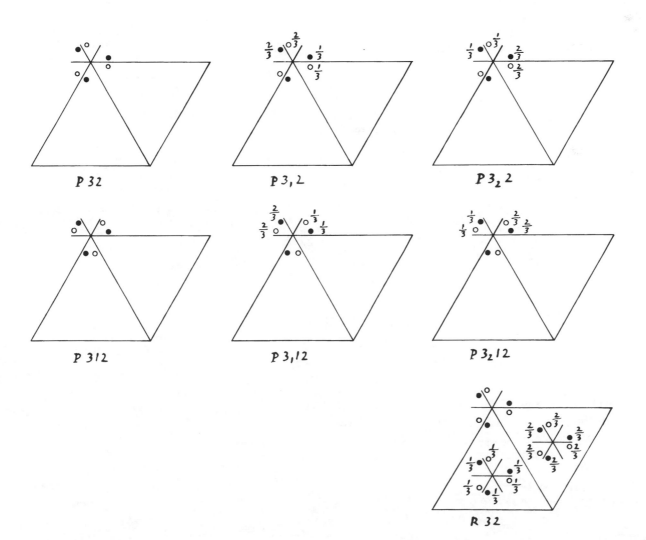

$P\,32$ $P\,3_1 2$ $P\,3_2 2$

$P\,312$ $P\,3_1 12$ $P\,3_2 12$

$R\,32$

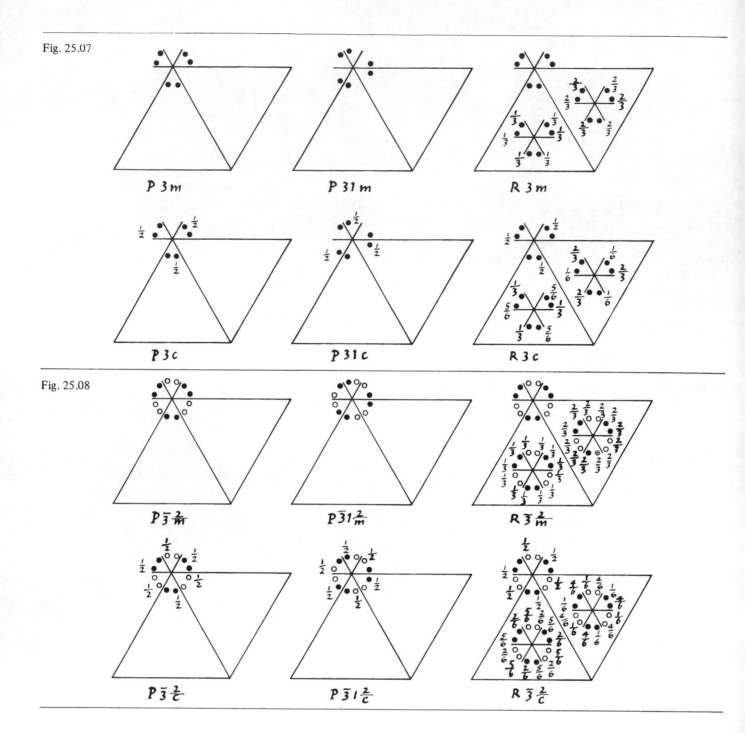

Fig. 25.07

P 3 m   P 31 m   R 3 m

P 3 c   P 31 c   R 3 c

Fig. 25.08

$P\bar{3}\frac{2}{m}$ $P\bar{3}1\frac{2}{m}$ $R\bar{3}\frac{2}{m}$

$P\bar{3}\frac{2}{c}$ $P\bar{3}1\frac{2}{c}$ $R\bar{3}\frac{2}{c}$

Point group 3m

This again divides into $3m$ and $31m$, but with the rhombohedral lattice (Fig. 25.01) only the former is possible. There are no screws, as they would be incompatible with the reflexions, and glides are possible only in the z-direction. There are thus six space groups:

$P3m$, $P31m$, $R3m$,
$P3c$, $P31c$, $R3c$.

These are shown in Fig. 25.07.

Point group $\bar{3}\frac{2}{m}$

The same remarks apply and there are again six space groups:

$P\bar{3}\frac{2}{m}$, $P\bar{3}1\frac{2}{m}$, $R\bar{3}\frac{2}{m}$,
$P\bar{3}\frac{2}{c}$, $P\bar{3}1\frac{2}{c}$, $R\bar{3}\frac{2}{c}$.

These are shown in Fig. 25.08.

The hexagonal system

As in the trigonal system, and for the same reasons, the only screws are about axes in the *z*-direction, and the only glides are *c*-glides.

Point group 6

There are six space groups, $P6$, $P6_1$ and $P6_5$, $P6_2$ and $P6_4$, $P6_3$. They are shown in Fig. 25.09.

Point group $\bar{6}$

There is only one space group, $P\bar{6}$ (Fig. 25.10).

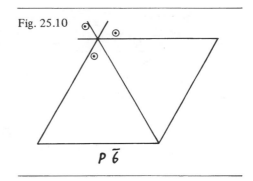

Fig. 25.10

$P\,\bar{6}$

Fig. 25.09

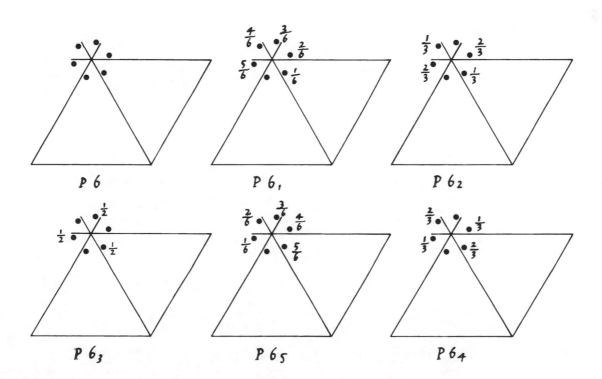

$P\,6$ $P\,6_1$ $P\,6_2$

$P\,6_3$ $P\,6_5$ $P\,6_4$

Point group 622

The hexad rotation can be replaced by screws, but not the diad. There are six space groups, as before, namely:

$P622$, $P6_1 22$ and $P6_5 22$, $P6_2 22$ and $P6_4 22$, $P6_3 22$.

They are shown in Fig. 25.11.

Point group $\frac{6}{m}$

The only screw allowing reflexion in the Z-plane is 6_3. There are therefore only two space groups, $6/m$ and $6_3/m$ (Fig. 25.12).

Fig. 25.11

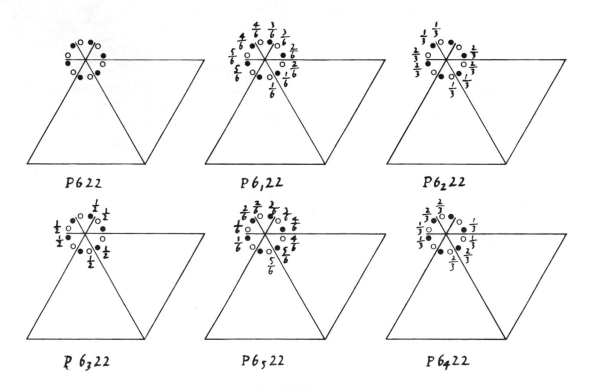

Fig. 25.12

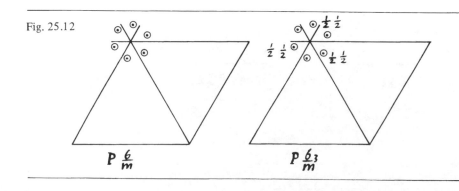

Point group $\bar{6}2m$ *(including* $\bar{6}m2$*)*
There are no screws and the only possible glides are in the *z*-direction. There are therefore four space groups:

$P\bar{6}2m$, $P\bar{6}2c$, $P\bar{6}m2$, $P\bar{6}c2$,

as shown in Fig. 25.13.

Fig. 25.13

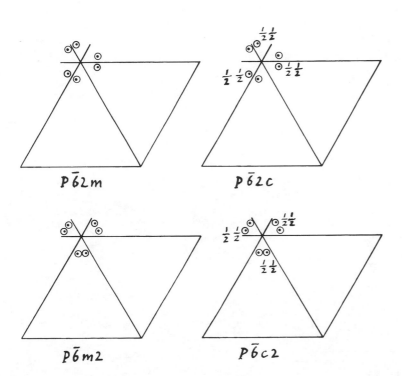

Fig. 25.14

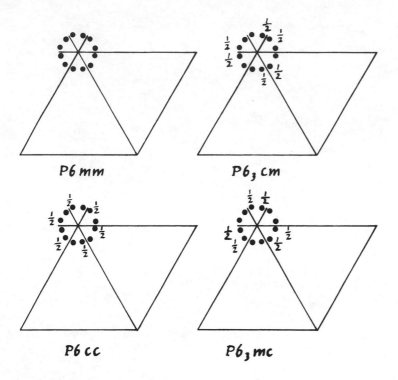

P6 mm

P6₃ cm

P6 cc

P6₃ mc

Fig. 25.15

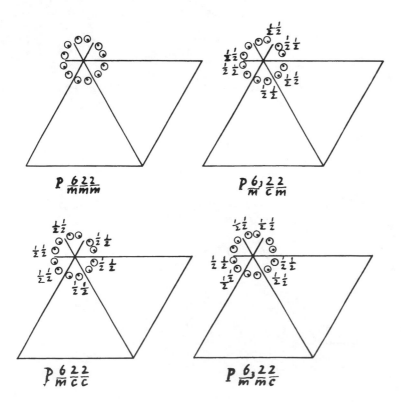

$P \frac{6}{m} \frac{2}{m} \frac{2}{m}$

$P \frac{6_3}{m} \frac{2}{c} \frac{2}{m}$

$P \frac{6}{m} \frac{2}{c} \frac{2}{c}$

$P \frac{6_3}{m} \frac{2}{m} \frac{2}{c}$

Point group 6mm

The rotation or screw is determined by the two reflexions or glides. The glides can only be in the z-direction. There are thus four space groups:

$P6mm$, $P6_3mc$, $P6_3cm$, $P6cc$,

as shown in Fig. 25.14.

Point group $\frac{6}{m}\frac{2}{m}\frac{2}{m}$

As glides are not possible in the Z-plane, we merely have to add m, in the first position, to the combinations of the last point group. Thus there are four space groups, based on

mmm, mmc, mcm, mcc.

The rotations and screws follow automatically (Fig. 25.15).

The cubic system

The symmetry of the lattice is $P\frac{4}{m}\bar{3}\frac{2}{m}$, where the first position after the prefix refers to axes in the three directions parallel to the edges of the cubes (and planes perpendicular to them), the second to axes along the four diagonals of the cubes, and the third to axes in the six directions parallel to diagonals of the faces (Fig. 25.16).

All the point groups in this system have at least the movements represented by the symbol 23, i.e. diad rotation about lines parallel to the edges and triad about the diagonals. In order to exhibit the triad rotation we use a modified form of diagram, showing only one-third of a set of equivalent points. Thus for the point group 23 we show four points (Fig. 25.17) related by the diad rotations. With origin and axes as shown, the coordinates of these points are

(x, y, z),　$(x, 1-y, -z)$,　$(1-x, y, -z)$,　$(1-x, 1-y, z)$,

but we write them as

(x, y, z),　$(x, -y, -z)$,　$(-x, y, -z)$,　$(-x, -y, z)$,

because it is understood that every point is repeated as unit intervals in all three directions. The movements from (x, y, z) to these four points are a subgroup of the 12 movements of the group. To complete the representation of the group we can add similar diagrams based on two more faces of the cube, as in Fig. 25.18(*a*). This shows a development of the three faces, the dots and rings in each face representing displacements perpendicular to that face. The patterns in the two new faces are identical with that in the original, but turned through a right angle. The three dots near the origin suggest the triad rotation symmetry about the diagonal $x = y = z$. The other three triad axes can be similarly suggested, as, for example, in Fig. 25.18(*b*). For our purposes it is usually sufficient to show the basic part of the diagram, as in Fig. 25.17. This illustrates the sub-group of movements. The other two squares, representing the cosets, are omitted.

Viewed along a diagonal of one of the cubes, the lattice is rhombohedral and has 'built-in' screw symmetry 3_1 and 3_2. (See Fig. 23.11, p. 000.) These movements 'pair' (or, rather, form a trio) with the triad rotation and will not provide any independent space groups.

Fig. 25.16

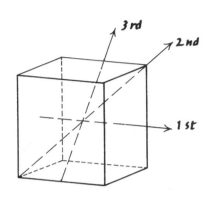

Fig. 25.17

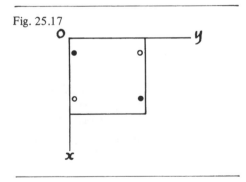

Fig. 25.18

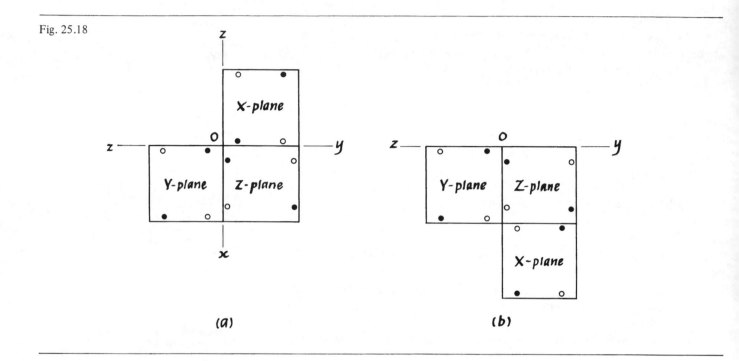

(a)   (b)

Point group 23

The point group has twelve members. With a primitive lattice we can have 23 or $2_1 3$, but with an *F*-lattice 2_1 pairs with 2 (Fig. 25.19). In either case the combinations with the translations give parallel axes at half-unit intervals. For $P2_1 3$ the screw axes in the three mutually perpendicular directions have to be non-intersecting and at $\frac{1}{4}$ unit distance apart, because intersecting 2_1 axes

Fig. 25.19

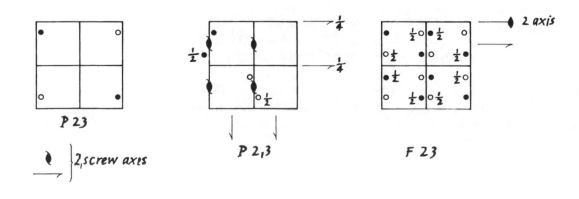

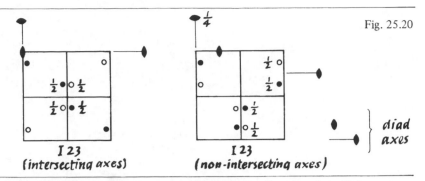

Fig. 25.20

I 23
(intersecting axes)

I 23
(non-intersecting axes)

diad axes

would imply a simple rotation in the perpendicular direction, but if the axes are $\frac{1}{4}$ unit apart they imply another screw of distance $\frac{1}{2}$ unit, as desired.*

With the *I*-lattice the rotation axes may be either intersecting or not (Fig. 25.20). In either case there are screw axes in intermediate planes.

Point group $\frac{2}{m}\bar{3}$

There are 24 movements in the group and of these, the 12 direct movements form the 23 group. So we can find the space groups by adding a reflexion or glide to the groups shown in Figs. 25.19, 25.20. The combination $2/m$ implies inversion and changes 3 to $\bar{3}$, provided the centre of inversion is on the triad axis. If the reflexion is replaced by a glide the centre of inversion is displaced by half the glide distance. Just as the 2 means that there are diad axes in three mutually perpendicular directions, so the *m* or *n* or *g* means reflexions or glides in three mutually perpendicular planes. Bearing these considerations in mind we can add *m* or *n* to *P*23; *a* to *P*$2_1$3; *m* or *d* to *F*23; and *m* or *g* to *I*23. These groups are shown in Fig. 25.21.

> *The truth of these last statements can be verified by examining the points in the diagram, but they can also be proved algebraically. (See Appendix 2.) The same applies to many other combinations of movements that occur in this and other chapters.

Fig. 25.21

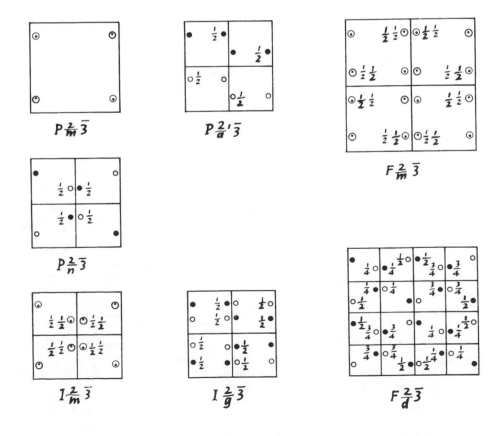

$P\frac{2}{m}\bar{3}$ $P\frac{2}{a}'\bar{3}$ $F\frac{2}{m}\bar{3}$

$P\frac{2}{n}\bar{3}$

$I\frac{2}{m}\bar{3}$ $I\frac{2}{g}\bar{3}$ $F\frac{2}{d}\bar{3}$

Point group 432

There are 24 movements in the group (p. 126) so for the diagram we need a sub-group of 8 members. This is conveniently taken as the one based on a single tetrad rotation and a diad, about axes at right angles (Fig. 25.22).

For space groups we note first that if there is a tetrad axis (rotation or screw) through the centre of a cell, the combinations with translations of the group give further such axes through the corners, and vice versa (Fig. 25.22). Secondly, diad rotations about axes in the diagonal direction are paired with screws about parallel axes (Fig. 25.23).

With the primitive lattice the tetrad rotations can be replaced by screws 4_1, 4_2 or 4_3, the 4_1 and 4_3 types being enantiomorphic. The arrangement of the screw axes is shown in Fig. 25.24. Note that these axes must themselves conform to the screw movements: that is why the axes are placed at distances apart equal to the displacement distances of the screws. (The 4_3 screw is of course a left-handed screw of distance $\frac{1}{4}$ unit). A set of three axes so related coincide with three non-intersecting edges of a cube, the diagonal of the cube through the two remaining vertices forming an axis of triad rotation. Thus it is seen that the triad axis always allows simple rotation, whether the tetrads are of simple rotation or screw. In the diagram for $P4_232$ there are two sets of

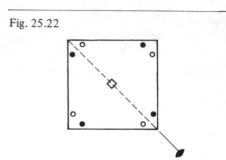

Fig. 25.22

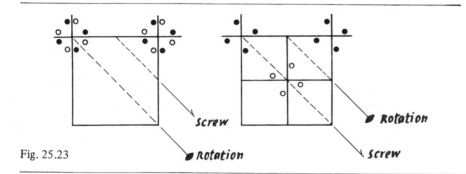

Fig. 25.23

Fig. 25.24

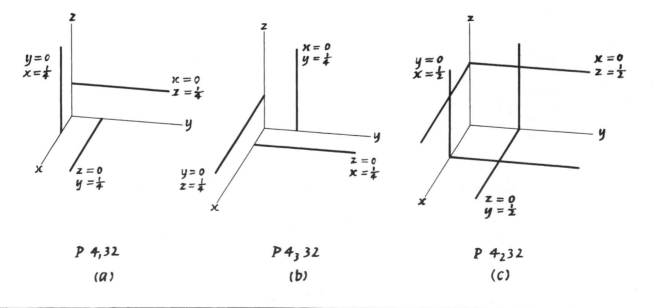

$P\,4_1 32$

(a)

$P\,4_3\,32$

(b)

$P\,4_2 32$

(c)

Fig. 25.25

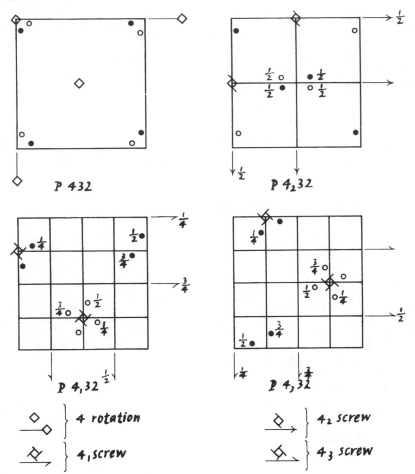

$P\,432$

$P\,4_2 32$

$P\,4_1 32\,\tfrac{1}{2}$

$P\,4_3 32$

◇◇ } 4 rotation

◇→ } 4_1 screw

◇→ } 4_2 screw

◇ } 4_3 screw

Fig. 25.26

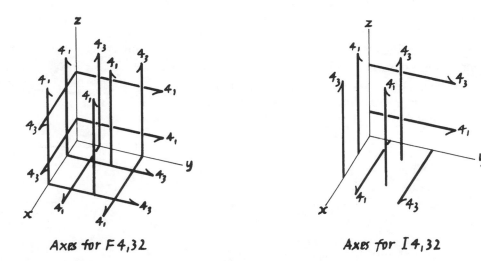

Axes for $F\,4_1 32$

Axes for $I\,4_1 32$

axes, to show the intersections. In the other two groups none of the screw axes intersect. Diagrams for the four space groups are shown in Fig. 25.25.

With an F-lattice or an I-lattice 4_2 is paired with 4 and 4_3 with 4_1. So we have only two space groups in each case. For the screw types the arrangement of the axes is shown in Fig. 25.26. In $F4_1 32$ the screw axes are repeated at half-unit intervals and hence intersect in pairs, though never three at a time. For $I4_1 32$ the axes are non-intersecting. Diagrams for these four groups are shown in Fig. 25.27.

Fig. 25.27

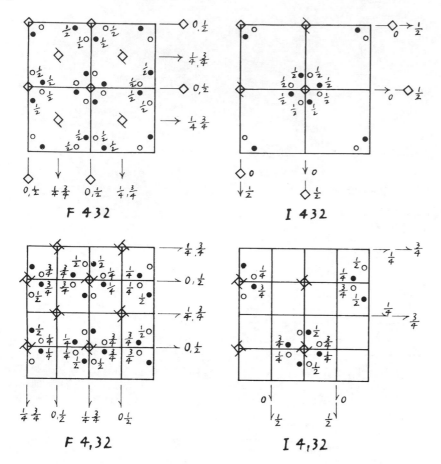

F 432

I 432

F 4₁32

I 4₁32

Point group $\bar{4}3m$

The point group contains 24 movements. (Apart from the identity, the three tetrad axes provide 9 movements, the four triads 8, and the six mirror-planes 6.) For the diagram we take the sub-group of 8 members based on a single $\bar{4}$ axis and a mirror plane containing it (Fig. 25.28).

If translations are added, to form the space group $P\bar{4}3m$, there will be parallel $\bar{4}$ axes through the corners of the squares, i.e. the $\bar{4}$ axes will be arranged as indicated for $P\bar{4}3m$ in Fig. 25.30. Inversion centres will occur at half-unit intervals along these axes. The axes themselves conform to the reflexions in the diagonal planes.

Variations are obtained by replacing m by g, c or n, but g pairs with m and c with n, as shown in Fig. 25.29. With a primitive lattice there are thus only two space groups. For the second one, $P\bar{4}3c$, the $\bar{4}$ axes, conforming to the glide reflexion, will no longer meet three at a time, but only two at a time, as they did for $P4_2 32$ (Fig. 25.24(c)). The inversion centres on the axes in the z-direction are at $(\frac{1}{4}, 0, \frac{1}{2})$, $(\frac{1}{4}, \frac{1}{2}, 0)$, $(\frac{3}{4}, 0, \frac{1}{2})$, $(\frac{3}{4}, \frac{1}{2}, 0)$.

With the F-lattice the pairings are the same and there are again only two groups, $F\bar{4}3m$ and $F\bar{4}3c$. With the I-lattice, however, g, c and n are each paired with m, but a diamond glide is possible. The groups are $I\bar{4}3m$ and $I\bar{4}3d$. For the latter, the axes are non-intersecting, as they were for $I4_1 32$ (Fig. 25.26) and the inversions centres are at such points as $(\frac{3}{8}, \frac{1}{4}, 0)$, $(\frac{7}{8}, \frac{1}{4}, 0)$, $(\frac{1}{8}, \frac{1}{4}, \frac{1}{2})$, etc.

Fig. 25.28

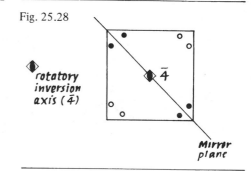

rotatory inversion axis ($\bar{4}$)

$\bar{4}$

Mirror plane

Fig. 25.29

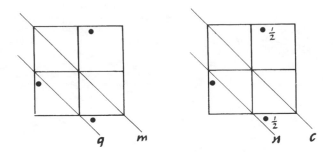

Fig. 25.30

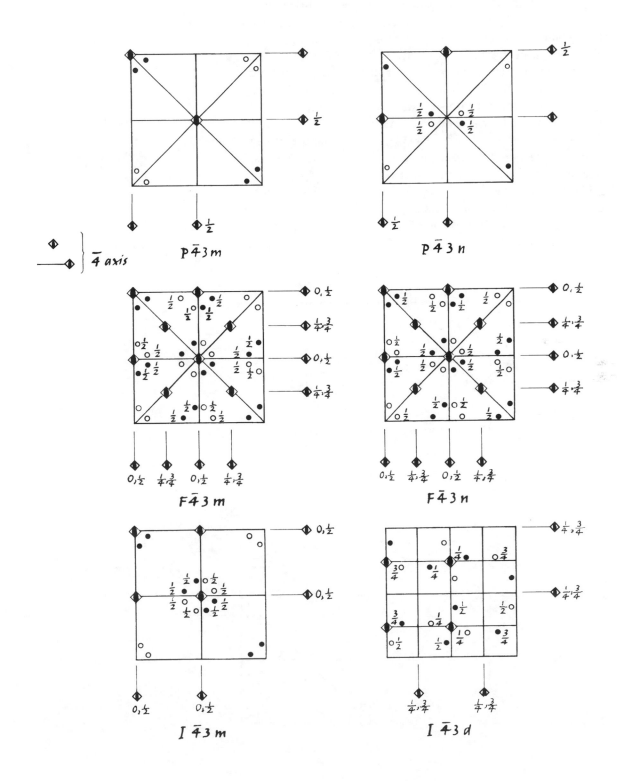

Point group $\frac{4}{m}\bar{3}\frac{2}{m}$

This is the full point group for the cube and contains 48 movements. The 24 direct movements are those of the point group 432 and the addition of one reflexion produces all the opposite movements. The space groups may therefore be found by adding a reflexion or glide to those of the point group 432. We choose a diagonal plane for this added movement, leaving the inversion and the reflexions or glides in other planes to follow automatically. As with the group $\bar{4}3m$, a *g*-glide in the diagonal plane pairs with a reflexion *m* in a parallel plane and a *c*-glide pairs similarly with an *n*-glide. With a primitive lattice there are therefore two space groups based on *P*432 and two based on *P*4$_2$32, as shown in Fig. 25.31. (There are none for *P*4$_1$32 or *P*4$_3$32, because if the system of axes for either of these groups, as shown in Fig. 25.24, is reflected in a diagonal plane the whole system becomes similar to that for *P*4$_2$32.)

With the *F*-lattice there are two space groups based on *F*432 and two on *F*4$_1$32. With the *I*-lattice the *n*-glide in the diagonal plane pairs with a simple reflexion, so there is only one space group based on *I*432. The disposition of the 4$_1$ and 4$_3$ axes for *I*4$_1$32, as shown in Fig. 25.26, requires that a space

---

Fig. 25.31

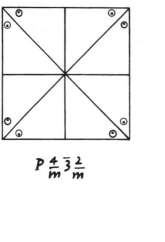

$$P\frac{4}{m}\bar{3}\frac{2}{m}$$

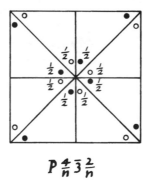

$$P\frac{4}{n}\bar{3}\frac{2}{n}$$

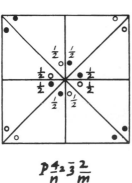

$$P\frac{4_2}{n}\bar{3}\frac{2}{m}$$

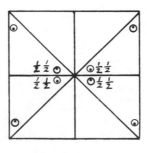

$$P\frac{4_2}{m}\bar{3}\frac{2}{n}$$

group can only be formed by having diamond glides in the diagonal planes and *g*-glides in planes parallel to sides of the cubes. These six space groups are shown in Fig. 25.32.

This completes the account of the 230 space groups. They are listed in Table 25.1.

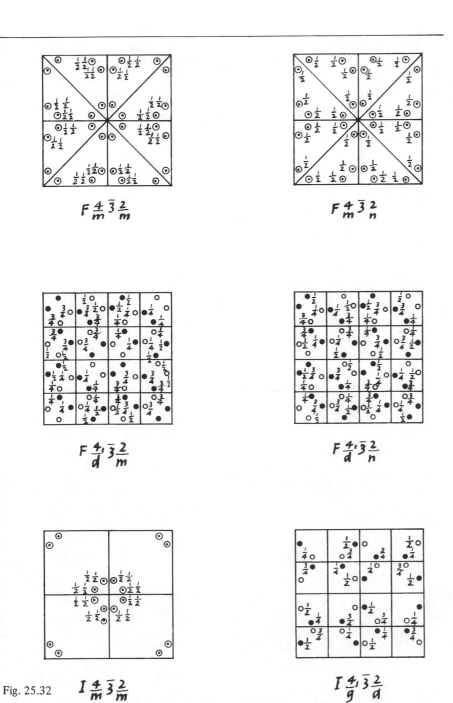

Fig. 25.32

$F\dfrac{4}{m}\bar{3}\dfrac{2}{m}$

$F\dfrac{4}{m}\bar{3}\dfrac{2}{n}$

$F\dfrac{4_1}{d}\bar{3}\dfrac{2}{m}$

$F\dfrac{4_1}{d}\bar{3}\dfrac{2}{n}$

$I\dfrac{4}{m}\bar{3}\dfrac{2}{m}$

$I\dfrac{4_1}{g}\bar{3}\dfrac{2}{d}$

Table 25.1 *The 230 space groups*

System	Point group	Space groups
Triclinic	1	$P1$
	$\bar{1}$	$P\bar{1}$
Monoclinic	2	$P2$ $P2_1$ $B2$
	m	Pm Pb Bm Bb
	$2/m$	$P2/m$ $P2_1/m$ $P2/b$ $P2_1/b$ $B2/m$ $B2/b$
Orthorhombic	222	$P222$ $P2_122$ $P22_12_1$ $P2_12_12_1$
		$C222$ $C2_122$ $F222$ $I222$ $I222$ (non-intersecting axes)
	$2mm$	$P2mm$ $P2_1mc$ $P2ma$ $P2_1mn$ $P2cc$ $P2_1ca$ $P2cn$
		$P2ba$ $P2_1bn$ $P2nn$ $C2mm$ C_12mc $C2cc$
		$A2mm$ $A2ma$ $A2bm$ $A2ba$ $I2mm$ $I2mc$ $I2cc$ $F2mm$ $F2dd$
	$\dfrac{2}{m}\dfrac{2}{m}\dfrac{2}{m}$	$P\dfrac{2}{m}\dfrac{2}{m}\dfrac{2}{m}$ $P\dfrac{2}{n}\dfrac{2_1}{m}\dfrac{2_1}{m}$ $P\dfrac{2}{m}\dfrac{2_1}{n}\dfrac{2_1}{n}$ $P\dfrac{2}{n}\dfrac{2}{n}\dfrac{2}{n}$ $P\dfrac{2}{a}\dfrac{2_1}{m}\dfrac{2}{m}$ $P\dfrac{2_1}{m}\dfrac{2_1}{n}\dfrac{2_1}{a}$
		$P\dfrac{2}{a}\dfrac{2}{n}\dfrac{2_1}{n}$ $P\dfrac{2}{m}\dfrac{2}{n}\dfrac{2_1}{c}$ $P\dfrac{2}{n}\dfrac{2_1}{c}\dfrac{2_1}{c}$ $P\dfrac{2}{m}\dfrac{2}{c}\dfrac{2}{c}$ $P\dfrac{2_1}{n}\dfrac{2}{c}\dfrac{2_1}{a}$ $P\dfrac{2_1}{m}\dfrac{2_1}{c}\dfrac{2}{a}$ $P\dfrac{2}{n}\dfrac{2}{b}\dfrac{2}{a}$
		$P\dfrac{2}{m}\dfrac{2_1}{b}\dfrac{2_1}{a}$ $P\dfrac{2}{a}\dfrac{2_1}{c}\dfrac{2}{c}$ $P\dfrac{2_1}{a}\dfrac{2_1}{b}\dfrac{2_1}{c}$
		$C\dfrac{2}{m}\dfrac{2}{m}\dfrac{2}{m}$ $C\dfrac{2_1}{m}\dfrac{2}{m}\dfrac{2}{c}$ $C\dfrac{2}{m}\dfrac{2}{c}\dfrac{2}{c}$ $C\dfrac{2}{a}\dfrac{2}{m}\dfrac{2}{m}$ $C\dfrac{2_1}{a}\dfrac{2}{m}\dfrac{2}{c}$ $C\dfrac{2}{a}\dfrac{2}{c}\dfrac{2}{c}$
		$I\dfrac{2}{m}\dfrac{2}{m}\dfrac{2}{m}$ $I\dfrac{2}{a}\dfrac{2}{m}\dfrac{2}{m}$ $I\dfrac{2}{m}\dfrac{2}{c}\dfrac{2}{c}$ $I\dfrac{2}{a}\dfrac{2}{c}\dfrac{2}{c}$ $F\dfrac{2}{m}\dfrac{2}{m}\dfrac{2}{m}$ $F\dfrac{2}{d}\dfrac{2}{d}\dfrac{2}{d}$
Tetragonal	4	$P4$ $P4_1$ $P4_3$ $P4_2$ $I4$ $I4_1$
	$\bar{4}$	$P\bar{4}$ $I\bar{4}$
	$4/m$	$P4/m$ $P4_2/m$ $P4/n$ $P4_2/n$ $I4/m$ $I4_1/g$
	422	$P422$ $P4_122$ $P4_222$ $P4_322$ $P42_12$ $P4_12_12$ $P4_22_12$
		$P4_32_12$ $I422$ $I4_122$
	$4mm$	$P4mm$ $P4gm$ $P4_2nm$ $P4_2cm$ $P4_2mc$ $P4_2gc$ $P4nc$ $P4cc$
		$I4mm$ $I4cm$ $I4_1md$ $I4_1cd$
	$\bar{4}2m$	$P\bar{4}2m$ $P\bar{4}2c$ $P\bar{4}2_1m$ $P\bar{4}2_1c$ $P\bar{4}m2$ $P\bar{4}g2$ $P\bar{4}c2$ $P\bar{4}n2$
		$I\bar{4}2m$ $I\bar{4}m2$ $I\bar{4}c2$ $I\bar{4}2d$
	$\dfrac{4}{m}\dfrac{2}{m}\dfrac{2}{m}$	$P\dfrac{4}{m}\dfrac{2}{m}\dfrac{2}{m}$ $P\dfrac{4_2}{m}\dfrac{2}{c}\dfrac{2}{m}$ $P\dfrac{4}{m}\dfrac{2}{g}\dfrac{2_1}{m}$ $P\dfrac{4_2}{m}\dfrac{2_1}{n}\dfrac{2}{m}$ $P\dfrac{4_2}{m}\dfrac{2}{m}\dfrac{2}{c}$ $P\dfrac{4}{m}\dfrac{2}{g}\dfrac{2}{c}$ $P\dfrac{4_2}{m}\dfrac{2_1}{g}\dfrac{2}{c}$
		$P\dfrac{4}{m}\dfrac{2_1}{n}\dfrac{2}{c}$ $P\dfrac{4_2}{n}\dfrac{2_1}{m}\dfrac{2}{m}$ $P\dfrac{4_2}{n}\dfrac{2_1}{c}\dfrac{2}{m}$ $P\dfrac{4}{n}\dfrac{2_1}{g}\dfrac{2}{m}$ $P\dfrac{4_2}{n}\dfrac{2}{n}\dfrac{2}{m}$ $P\dfrac{4_2}{n}\dfrac{2_1}{m}\dfrac{2}{c}$ $P\dfrac{4}{n}\dfrac{2_1}{c}\dfrac{2}{c}$
		$P\dfrac{4_2}{n}\dfrac{2}{g}\dfrac{2}{c}$ $P\dfrac{4}{n}\dfrac{2}{n}\dfrac{2}{c}$ $I\dfrac{4}{m}\dfrac{2}{m}\dfrac{2}{m}$ $I\dfrac{4}{m}\dfrac{2}{c}\dfrac{2}{m}$ $I\dfrac{4_1}{g}\dfrac{2}{m}\dfrac{2}{d}$ $I\dfrac{4_1}{g}\dfrac{2}{c}\dfrac{2}{d}$
Trigonal	3	$P3$ $P3_1$ $P3_2$ $R3$
	$\bar{3}$	$P\bar{3}$ $R\bar{3}$
	32	$P32$ $P3_12$ $P3_22$ $P312$ $P3_112$ $P3_212$ $R32$
	$3m$	$P3m$ $P31m$ $R3m$ $P3c$ $P31c$ $R3c$
	$\bar{3}2/m$	$P\bar{3}\dfrac{2}{m}$ $P\bar{3}1\dfrac{2}{m}$ $R\bar{3}\dfrac{2}{m}$ $P\bar{3}\dfrac{2}{c}$ $P\bar{3}1\dfrac{2}{c}$ $R\bar{3}\dfrac{2}{c}$
Hexagonal	6	$P6$ $P6_1$ $P6_5$ $P6_2$ $P6_4$ $P6_3$
	$\bar{6}$	$P\bar{6}$
	622	$P622$ $P6_122$ $P6_522$ $P6_222$ $P6_422$ $P6_322$
	$6/m$	$P6/m$ $P6_3/m$
	$\bar{6}2m$	$P\bar{6}2m$ $P\bar{6}2c$ $P\bar{6}m2$ $P\bar{6}c2$
	$6mm$	$P6mm$ $P6_3cm$ $P6_3mc$ $P6cc$
	$\dfrac{6}{m}\dfrac{2}{m}\dfrac{2}{m}$	$P\dfrac{6}{m}\dfrac{2}{m}\dfrac{2}{m}$ $P\dfrac{6_3}{m}\dfrac{2}{c}\dfrac{2}{m}$ $P\dfrac{6_3}{m}\dfrac{2}{m}\dfrac{2}{c}$ $P\dfrac{6}{m}\dfrac{2}{c}\dfrac{2}{c}$
Cubic	23	$P23$ $P2_13$ $F23$ $I23$ (intersecting axes)
		$I23$ (non-intersecting axes)
	$\dfrac{2}{m}\bar{3}$	$P\dfrac{2}{m}\bar{3}$ $P\dfrac{2}{n}\bar{3}$ $P\dfrac{2_1}{a}\bar{3}$ $F\dfrac{2}{m}\bar{3}$ $F\dfrac{2}{d}\bar{3}$ $I\dfrac{2}{m}\bar{3}$ $I\dfrac{2}{g}\bar{3}$
	432	$P432$ $P4_232$ $P4_132$ $P4_332$
		$F432$ $F4_132$ $I432$ $I4_132$
	$\bar{4}3m$	$P\bar{4}3m$ $P\bar{4}3c$ $F\bar{4}3m$ $F\bar{4}3c$ $I\bar{4}3m$ $I\bar{4}3d$
	$\dfrac{4}{m}\bar{3}\dfrac{2}{m}$	$P\dfrac{4}{m}\bar{3}\dfrac{2}{m}$ $P\dfrac{4}{n}\bar{3}\dfrac{2}{n}$ $P\dfrac{4_2}{m}\bar{3}\dfrac{2}{n}$ $P\dfrac{4_2}{n}\bar{3}\dfrac{2}{m}$
		$F\dfrac{4}{m}\bar{3}\dfrac{2}{m}$ $F\dfrac{4}{m}\bar{3}\dfrac{2}{n}$ $F\dfrac{4_1}{d}\bar{3}\dfrac{2}{m}$ $F\dfrac{4_1}{d}\bar{3}\dfrac{2}{n}$ $I\dfrac{4}{m}\bar{3}\dfrac{2}{m}$ $I\dfrac{4_1}{g}\bar{3}\dfrac{2}{d}$

26

Limiting groups

Two dimensions

In two dimensions there are two series of point groups

   1, 2,   3, 4,   5, ... ;

and 1*m*, 2*mm*, 3*m*, 4*mm*, 5*m*, ...

As the rotation number increases the groups approach the limits ∞ and ∞m. From the geometrical point of view these are the same. In Fig. 26.01, which shows plan views of two kinds of toothed wheel, a large increase in the number of teeth brings both kinds towards the same limit, namely a smooth cylinder, seen in plan as a circle. The lack of reflexional symmetry in Fig. 26.01(*a*) disappears as the limit is reached and, for geometrical purposes, the limit for both series is ∞m. But it is possible to have physical illustrations of the limiting group ∞, for example a disc round which there is a field of magnetic force (Fig. 26.02), or a crystal that rotates the plane of polarized light. A rotating disc provides another illustration, but it is necessary to distinguish between the one-way physical rotation of the disc and the imaginary rotations by which we describe the symmetry. The latter are movements of a group and are necessarily both clockwise and anticlockwise. The traffic roundabout shown in Fig. 26.03 has tetrad symmetry, i.e. it would be unchanged if rotated through a right angle either way. But it has no reflexional symmetry because the traffic flow is one-way.

In this book we deal only with geometrical symmetry and on that basis we say that there is only one continuous two-dimensional point group, ∞m.

In a somewhat similar way the friezes in Fig. 26.04, of types *r*11*m*, *r*2*mm* and *r*1, become continuous bands if the translation interval is reduced to zero. Geometrically the limits of the first two are the same, and that of the third differs from them in the only possible way, i.e. in not allowing reflexion in a

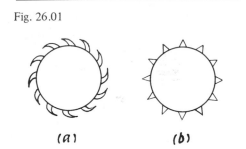

Fig. 26.01

(*a*)   (*b*)

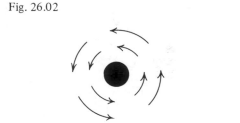

Fig. 26.02

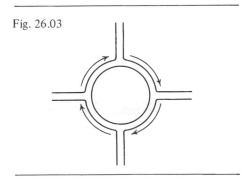

Fig. 26.03

Fig. 26.04

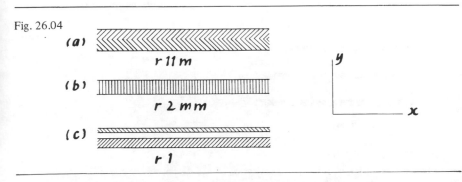

(*a*) *r 11 m*

(*b*) *r 2 m m*

(*c*) *r 1*

longitudinal mirror line. These two limiting types are shown in Fig. 26.05, where the prefix r_0 indicates that the translation interval has been reduced to zero. It will be noticed that the two types correspond to the two one-dimensional point groups.

In a 'wallpaper' group the translations are represented by a net and the translation interval may be reduced to zero either in one direction or in all directions, the net becoming a set of parallel lines or a uniform continuous plane. Such groups are called *semi-continuous* and *continuous* respectively. They will be denoted by the prefixes p_0 and p_{00}.

If the interval is reduced in one direction only, the cross-section is a one-dimensional frieze and there are again two types, as shown in Fig. 26.06. If it is reduced in two, and hence in all directions in the plane, the pattern becomes a uniform plane. The crystallographic restriction disappears and any rotation is possible. The symbol is $p_{00}\infty m$.

Three dimensions

In a point group, only rotations can be continuous. A three-dimensional point group can include continuous rotations either about one axis or about all axes (for continuous rotations about diameters OA, OB of a sphere imply, by a limiting case of Euler's Construction, a continuous rotation about any other diameter in the plane OAB; moreover a rotation about a given diameter OP outside that plane can be obtained by combining suitable rotations about any two diameters in the plane, the angle of rotation being varied continuously as the two chosen diameters vary.)

The groups that include rotations about one axis lead to two limiting cases according to whether there is or is not reflexion in a plane perpendicular to that axis. These limiting groups are ∞m and $\frac{\infty}{m}\frac{2}{m}$, represented respectively by the circular cone and the circular cylinder. If there is a rotation about all axes through the point, the group is $\frac{\infty}{m}\frac{\infty}{m}$, represented by the sphere.

Fig. 26.05

Fig. 26.06

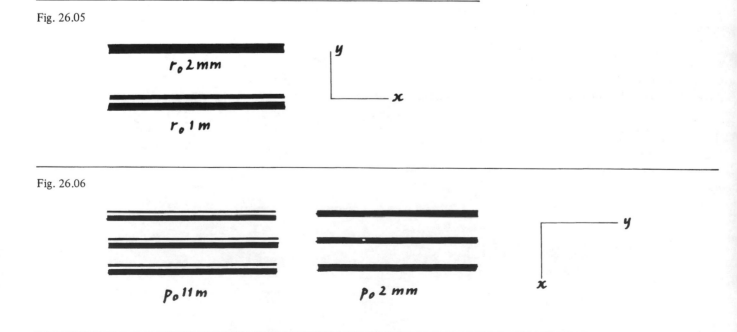

Three-dimensional groups that include translations in one direction ('rod' groups) are formed by combining movements of the translation group with those of a point group. Either of these or both may be continuous. We consider first the groups in which the point group is continuous but not the translations. The first two of the three point groups just mentioned can be used, giving the rod groups $r\infty m$ and $r\frac{\infty}{m}\frac{2}{m}$ (Fig. 26.07).

If a discontinuous point group is combined with a continuous translation the geometrical representation is a prism of infinite length, $r_0 n$ or $r_0 nm$ (n odd) or $r_0 nmm$ (n even), n being a positive integer. Typical cross-sections are as in Fig. 26.01. If both point group and translation are continuous there is only one group $r_0\frac{\infty}{m}\frac{2}{m}$, represented by the circular cylinder.

Finally we may replace the continuous rotation by a continuous screw. It will be recalled that the screws 4_2 and 6_2 include diad rotation, and that 6_3 includes triad. In the same way a continuous screw may allow a discontinuous simple rotation. This is illustrated, for example, by a four-threaded screw (Fig. 26.08) which has tetrad rotation for any cross-section and thus for the screw as a whole. As a symbol for this, analogous to 4_2 and 6_3, we use ∞_4, with a superscript $+$ or $-$ to indicate whether the screw is right-handed or left-handed. The symmetry group for the screw illustrated is thus $r\infty_4^+2$. The diad rotation is demonstrated by the fact that a nut fitting such a screw could be placed on it either way round. There are two infinite sets of such groups, $r\infty_n^+2$ and $r\infty_n^-2$, where n is any positive integer. The value $n = 1$ corresponds to the ordinary single-threaded screw.

Three-dimensional groups that include translations in two directions ('layer' groups) may be continuous or semi-continuous. The continuous ones, i.e. with continuous translations in all directions in a plane, are of two types $p_{00}\frac{\infty}{m}\frac{2}{m}$ and $p_{00}\infty m$. They are illustrated by parallel planes or layers, with or without reflexion, with cross-sections as shown in Fig. 26.05.

Fig. 26.07

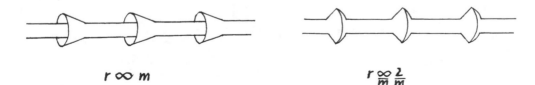

$r \infty m$ $r \frac{\infty}{m} \frac{2}{m}$

Fig. 26.08

For a semi-continuous layer pattern the cross-section may be any one of the seven friezes. Taking the z-axis perpendicular to the layer, and the y-axis in the direction of the continuous translation, the seven groups are:

$$p_0 11m, \quad p_0 11\tfrac{2}{m}, \quad p_0 2mm, \quad p_0 m2m, \quad p_0 \tfrac{2}{m}\tfrac{2}{m}\tfrac{2}{m}, \quad p_0 a2_1 m, \quad p_0 \tfrac{2}{a}\tfrac{2_1}{m}\tfrac{2}{m}.$$

They are illustrated in Fig. 26.09.

Lastly we consider three-dimensional space groups. These may include continuous translations in one direction only, or over a plane, or in all directions in space. If in one direction only, the cross-section of the pattern is one of the 17 'wallpapers'. The symbols for these groups are:

$$P_0 1, \quad P_0 2, \quad P_0 3, \quad P_0 4, \quad P_0 6, \quad P_0 1g, \quad P_0 2mg, P_0 2gg, \quad P_0 4gm,$$
$$P_0 1m, P_0 2mm, P_0 3m1, P_0 31m, P_0 4mm, P_0 6mm, \; C_0 1m, \; C_0 2mm.$$

With continuity over a plane the cross-section is a semi-continuous wallpaper and there are two types, as shown in Fig. 26.06. The group symbols are $P_{00}\infty m$ and $P_{00}\tfrac{\infty}{m}\tfrac{2}{m}$. Last of all there is the possibility of continuity in all directions in space and the only geometrical representation of this is space itself. If a symbol were required it would be $P_{000}\tfrac{\infty}{m}\tfrac{\infty}{m}\tfrac{\infty}{m}$.

Fig. 26.09

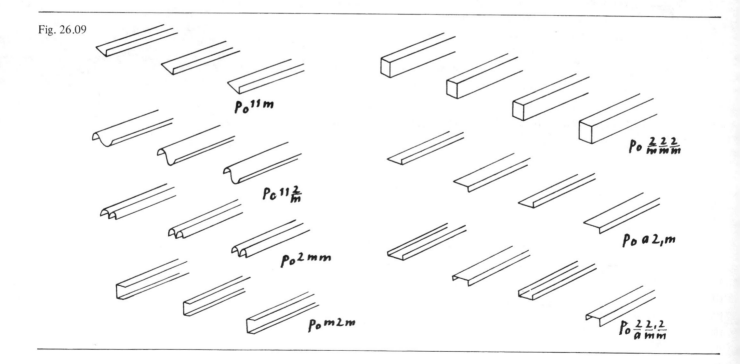

27

Colour symmetry

The introduction of colour into a symmetrical pattern has an effect very like the addition of an extra dimension. It is indeed possible to represent the colours in that way and we shall sometimes find it convenient to do so. But it must be emphasized that the coloured pattern exists in its own right and the extra dimension is only a means of representation. To envisage diagrams or models in which the extra dimension occurs as well as the colours, is to confuse the issue. There should be one or the other.

A pattern is said to have colour symmetry if a change from either of two colours to the other, or a step in a colour sequence, occurs in conjunction with one or more of its symmetry movements; and a colour symmetry group is one that contains movements carrying such changes of colour. For example, the two-coloured point group 4′ (Fig. 27.01(a)) consists of

> the identity,
> rotation through two right angles,
> coloured rotations through one right angle and three right angles.

Similarly the three-coloured group $6^{(3)}$ (Fig. 27.01(b)) consists of six movements, namely

> the identity,
> a rotation through two right angles,
> rotations of 60° and 240° accompanied by one step of the colour
>   sequence,
> rotations of 120° and 300° accompanied by two steps of the colour
>   sequence.

Dichromatic symmetry groups

We speak in terms of colour as a matter of convenience, but it is understood that two colours can represent any kind of polarity. If, as suggested in Part I, two colours can be represented by + and −, or by front and back, it is equally true, as any electrician knows, that + and − can be represented by colours. And we do not use such terms as *antisymmetry* and *antirotation*, because they seem inappropriate and can be easily, if less compactly, replaced by *dichromatic symmetry* and *dichromatic rotation*. The idea of colour serves all purposes.

For notation, if a movement is accompanied by a change of colour, a prime is added to the symbol. Thus while 4 represents a quarter-turn, 4′ is a quarter-turn with change of colour. On the same principle 1′ is the identity with change of colour, applicable to a pattern consisting of 'grey' or neutral elements (1′ is usually placed at the end of a symbol; but it should be noted that 3′, 5′, etc., include 1′). Examples are shown in Fig. 27.02.

The symbols for repetition are *r* for a row, *p* or *c* for a net, and *P, A, B, C, I* or *F* for a lattice. Primes will be used with these also, but in two dimensions or

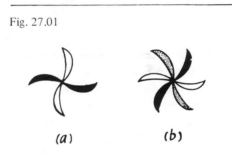

Fig. 27.01

(a)   (b)

Fig. 27.02

Fig. 27.03

in three it will be necessary to use a suffix to indicate the direction in which the change of colour occurs. In two dimensions we use a and b for the directions of the axes of x and y respectively, and n for that of the diagonal. In three dimensions we use suffixes a, b, c for the x-, y- and z-directions respectively, and capitals A, B, C for the diagonals of the planes perpendicular to them, the X-, Y- and Z-planes.

The representation of colours by means of an extra dimension provides an alternative form of symbol. We shall give this form in parentheses.

(a)   (b)

One dimension
Point groups

Without colours there were only two point groups, namely those with and without reflexion (Fig. 27.03). As it is difficult to show black and white and grey in one dimension we use the '+ and −' representation for the coloured groups (Fig. 27.04(a)); but these five coloured point groups appear even more clearly if we use the extra dimension (Fig. 27.04(b)).

Line groups

There are two one-dimensional line groups, formed from the point groups of Fig. 27.03, and they give seven coloured groups (Fig. 27.05(a)). It will be noticed that in the last pair the alternation of + and − is associated with the translation. If we use the extra dimension we have Fig. 27.05(b), which will be recognized at once as representing the seven uncoloured friezes.

Two dimensions
Point groups

There are an infinite number of uncoloured point groups, ten of them being 'crystallographic'. With two colours it is possible to use the third dimension to represent change of colour. For those who wish to adopt that method it may be noted that any of the 36 point groups listed on p. 131 may be used, except the five polyhedral groups in the last line. Ten of them (those in the third and fourth lines, together with $\bar{6}$ and $\bar{6}m2$, corresponding to the blank spaces in those lines) represent the 'grey' or neutral types. Thus there are 10 single-coloured groups, 10 'grey' ones and 11 particoloured.

We prefer, however, to find the particoloured groups by using the fact that in any such group the movements not involving change of colour form a subgroup of index 2, i.e. containing half the members of the group.

Proof. If C_i ($i = 1, 2, 3, \ldots, n$) are the movements involving change of colour and A_j ($j = 1, 2, 3, \ldots, m$) those without, we have $C_i A_j = C_k$, where C_k is one of the movements C_i. For a given C_i, as j takes m different values, there are m different C_k (for if $C_i A_j = C_i A_h$, pre-multiplication by C_i^{-1} shows that $A_j = A_h$). Therefore $n \geqslant m$. Similarly $C_i C_j = A_k$ and $m \geqslant n$. Therefore $m = n$. We therefore look for sub-groups of index 2 in each of the 10 crystallographic point groups. For example, in the point group $4mm$ there are two such sub-groups, giving two particoloured groups, as shown in Fig. 27.06. The complete list is as given in Table 27.1.

Fig. 27.04

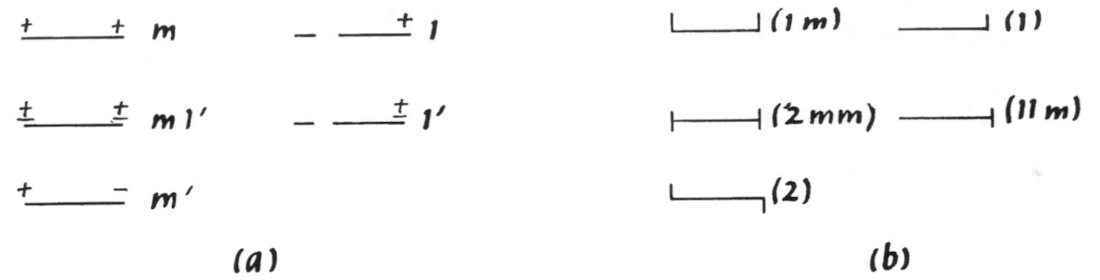

$+ \quad \pm \ m$ $\quad - \ \ + \ 1$ $\quad$ (1m) $\quad$ (1)

$\pm \quad \pm \ m1'$ $\quad - \ \ \pm \ 1'$ $\quad$ (2mm) $\quad$ (11m)

$+ \quad - \ m'$ $\quad$ (2)

(a) (b)

.Fig. 27.05

$+\ +\ +\ +\ +\ +\ +\ +\ \ rm$ $\quad -\ +\ -\ +\ -\ +\ -\ +\ \ r1$

$\pm\ \pm\ \pm\ \pm\ \pm\ \pm\ \pm\ \pm\ \ rm1'$ $\quad -\ \pm\ -\ \pm\ -\ \pm\ -\ \pm\ \ r1'$

$+\ -\ +\ -\ +\ -\ +\ -\ \ rm'$

$+\ +\ -\ -\ +\ +\ -\ -\ \ r'm$ $\quad -\ +\ -\ -\ -\ \pm\ -\ -\ \ r'1$

(a)

(r1m) (r1)

(r2mm) (r11m)

(r2)

(r2mg) (r11g)

(b)

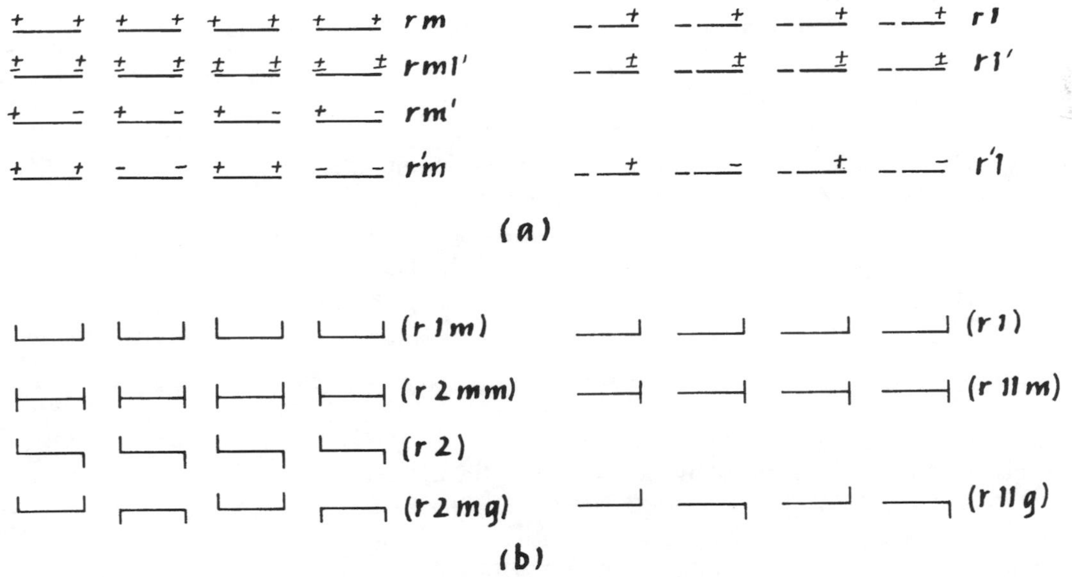

Fig. 27.06

Group	Sub-group	Particoloured group

$4mm$ 4 $4m'm'$

$2mm$ $4'mm'$

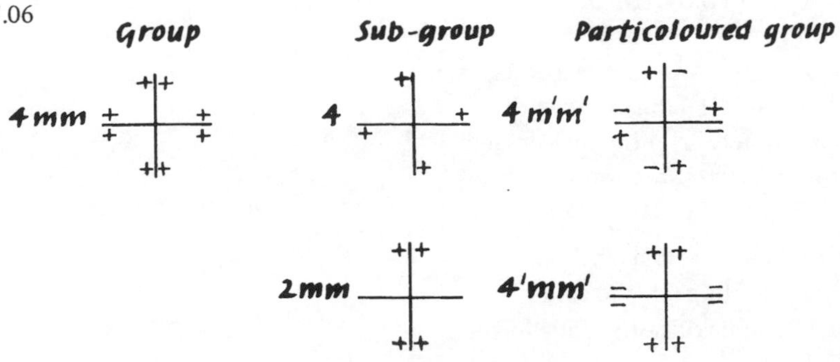

Table 27.1

Group	Sub-group	Particoloured group	
2	1	$2'$	$(\bar{1})$
4	2	$4'$	$(\bar{4})$
6	3	$6'$	$(\bar{3})$
$1m$	1	$1m'$	(12)
$2mm$	2	$2m'm'$	(222)
	$1m$	$2'm'm$	$(\bar{1}\frac{2}{m})$
$3m$	3	$3m'$	(32)
$4mm$	4	$4m'm'$	(422)
	$2mm$	$4'm'm$	$(\bar{4}2m)$
$6mm$	6	$6m'm'$	(622)
	$3m$	$6'm'm$	$(\bar{3}\frac{2}{m})$

Line groups

For line groups the natural method is to work from the seven uncoloured frieze patterns, applying black and white, or alternatively + and − signs, to indicate the polarity. Each of the seven gives a single-coloured group, a grey one, and one in which the colour changes with the translation. In addition there are ten more in which there is a colour change within the point group, making 31 in all, as shown in Fig. 27.07.

If we use a third dimension, i.e. if white and black are taken to mean front and back, we have the 31 'ribbon ornaments' referred to in the footnote on p. 133. Another way to derive these 31 groups is, as suggested in that footnote, to select from the 75 three-dimensional line groups ('rod groups') the 22 that contain no more than diad rotation, with 9 more obtained by interchanging the x- and y-axes. (If this is done, however, it must be borne in mind that in the rod groups the z-axis is taken in the direction of the translation, whereas in the frieze groups it is perpendicular to the plane, with the x-axis in the translation direction.)

Plane groups

If the dichromatic plane groups are represented by means of the extra dimension we have the 80 three-dimensional plane groups described in Chapter 22. This provides one way of enumerating the coloured groups but it brings them out in a somewhat unnatural order and we therefore prefer to list them as derived from the 17 'wallpaper' groups. To make up the total of 80 we have 17 single-coloured groups, 17 'grey' ones and hence 46 particoloured. Of these last, some are formed by alternation of colour with rotations or reflexions or glides and some by changing the colour with one or other of the translations, according to the black-and-white nets shown in Fig. 11.11 (p. 64). Those of the first kind have the prefix p or c for an uncoloured net, primitive or centred, and those of the second kind p'_a, p'_b or p'_n, according to the direction in which the change of colour occurs.

In Fig. 27.09 we take the wallpaper groups in the same order as in Fig. 18.05 (p. 118) and for each we give the two kinds of derived groups in the order mentioned. There is no need to consider changes of colour of both kinds in the

Fig. 27.07

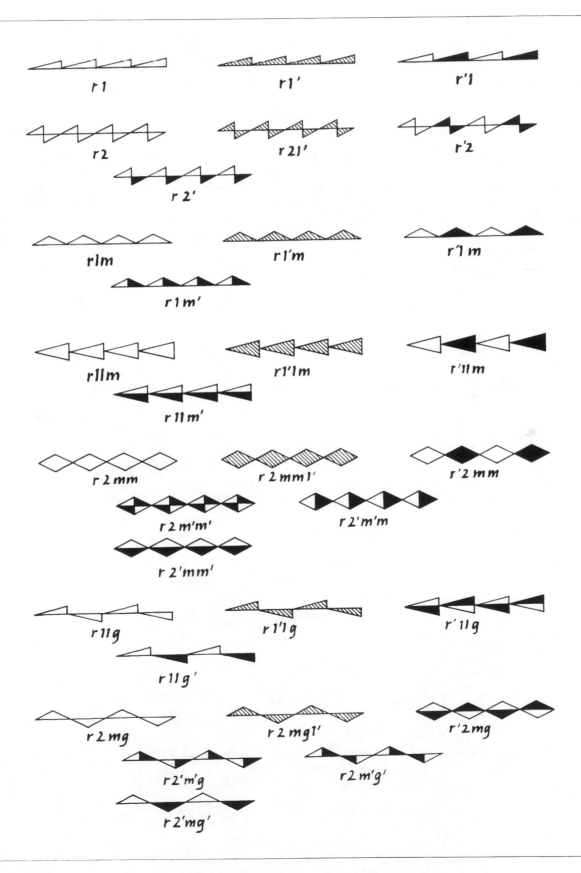

Fig. 27.09

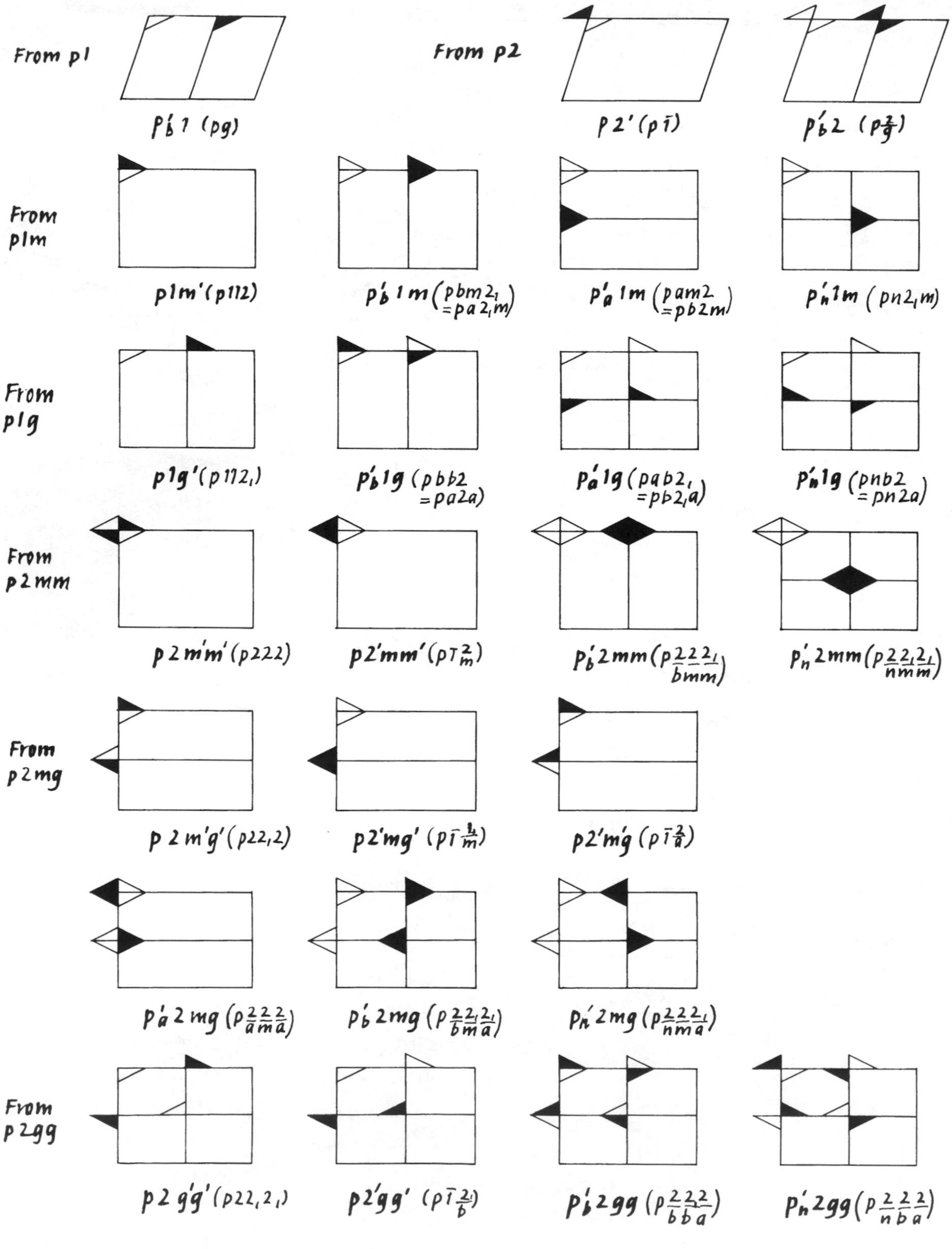

From p1

$p_b' 1$ (pg)

From p2

$p2'$ ($p\bar{1}$)

$p_b' 2$ ($p\frac{2}{g}$)

From p1m

$p1m'$ ($p112$) $p_b'1m$ $\left(\begin{array}{c}pbm2_1\\=pa2_1m\end{array}\right)$ $p_a'1m$ $\left(\begin{array}{c}pam2\\=pb2m\end{array}\right)$ $p_n'1m$ ($pn2_1m$)

From p1g

$p1g'$ ($p112_1$) $p_b'1g$ ($pbb2$ $=pa2a$) $p_a'1g$ ($pab2_1$ $=pb2_1a$) $p_n'1g$ ($pnb2$ $=pn2a$)

From p2mm

$p2m'm'$ ($p222$) $p2'mm'$ ($pT\frac{2}{m}$) $p_b'2mm$ ($p\frac{2}{b}\frac{2}{m}\frac{2_1}{m}$) $p_n'2mm$ ($p\frac{2}{n}\frac{2_1}{m}\frac{2_1}{m}$)

From p2mg

$p2m'g'$ ($p22_12$) $p2'mg'$ ($p\bar{1}\frac{2_1}{m}$) $p2'm'g$ ($p\bar{1}\frac{2}{a}$)

$p_a'2mg$ ($p\frac{2}{a}\frac{2}{m}\frac{2}{a}$) $p_b'2mg$ ($p\frac{2}{b}\frac{2_1}{m}\frac{2_1}{a}$) $p_n'2mg$ ($p\frac{2}{n}\frac{2}{m}\frac{2_1}{a}$)

From p2gg

$p2g'g'$ ($p22_12_1$) $p2'gg'$ ($p\bar{1}\frac{2_1}{b}$) $p_b'2gg$ ($p\frac{2}{b}\frac{2}{b}\frac{2}{a}$) $p_n'2gg$ ($p\frac{2}{n}\frac{2}{b}\frac{2}{a}$)

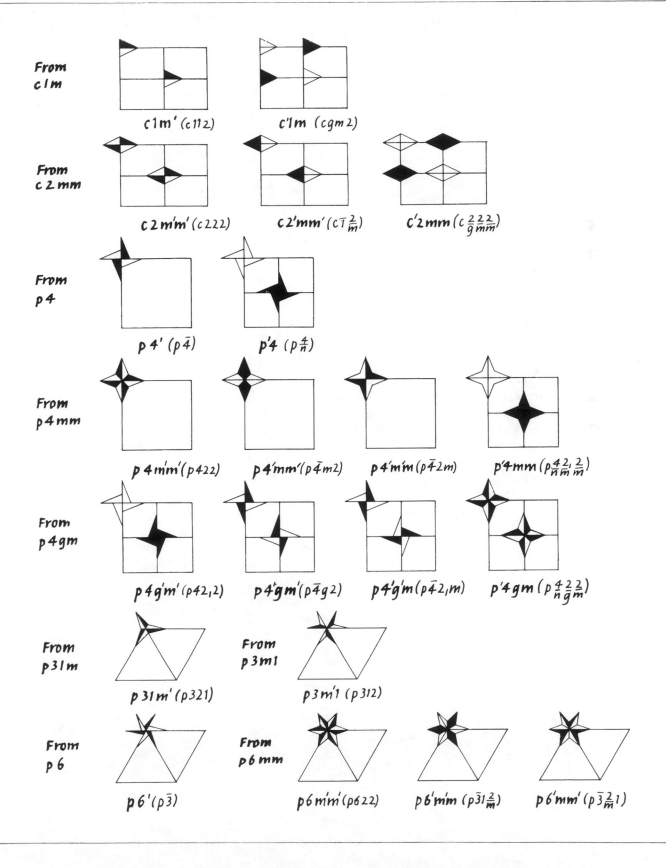

From
c l m

$c\,1m'\,(c\,112)$ $c'1m\,(c\,gm\,2)$

From
c 2 mm

$c\,2\,m'm'\,(c\,222)$ $c\,2'mm'\,(c\,1\tfrac{2}{m})$ $c'2mm\,(c\,\tfrac{2}{g}\tfrac{2}{m}\tfrac{2}{m})$

From
p 4

$p\,4'\,(p\,\bar{4})$ $p'4\,(p\,\tfrac{4}{n})$

From
p 4 mm

$p\,4\,m'm'\,(p\,422)$ $p\,4'mm'\,(p\,\bar{4}m2)$ $p\,4'm'm\,(p\,\bar{4}2m)$ $p'4mm\,(p\,\tfrac{4}{n}\tfrac{2}{m}\tfrac{2}{m})$

From
p 4gm

$p\,4\,g'm'\,(p\,42_12)$ $p\,4'g'm\,(p\,\bar{4}g2)$ $p\,4'g'm\,(p\,\bar{4}2_1m)$ $p'4gm\,(p\,\tfrac{4}{n}\tfrac{2}{g}\tfrac{2}{m})$

From
p 31m

$p\,31m'\,(p\,321)$

From
p 3m1

$p\,3m'1\,(p\,312)$

From
p 6

$p\,6'\,(p\,\bar{3})$

From
p 6mm

$p\,6\,m'm'\,(p\,622)$ $p\,6'm'm\,(p\,\bar{3}1\tfrac{2}{m})$ $p\,6'mm'\,(p\,\bar{3}\tfrac{2}{m}1)$

same derived group. For suppose that a white feature A of a pattern is transformed to a black one X by a rotation, reflexion or glide and to another black one Y by a translation: then X and Y are necessarily related and can be regarded as part of a pattern of the second kind. Thus, for example, $p_b' 1m = p_b' 1g$ (Fig. 27.08(*a*)) and $p_n' 2' = p_n' 2$ (Fig. 27.08(*b*)).

In the symbols we do not give the reflexions or glides in intermediate lines, as they do not help in distinguishing the different types. It is to be noted that reflexions or glides in two different directions determine the rotation, so in such cases the numeral is not strictly necessary, but we give it because it helps to characterize the group. The symbols for the three-dimensional representation are given as before in parentheses.

Three dimensions

A dichromatic three-dimensional group can of course be represented by an uncoloured group in four dimensions. But a more practicable method for enumerating such groups is to apply + and −, or black and white, to the corresponding uncoloured patterns.

Point groups

There are an infinite number of uncoloured point groups, but we restrict ourselves to the 32 'crystallographic' ones. With two colours available there are thus 32 single-coloured (polar) groups, and 32 'grey' (neutral) ones. In addition there are 58 particoloured groups (of mixed polarity). To find these we recall that in any such group the movements not involving change of colour form a sub-group of index 2, i.e. one containing half the members of the group. We therefore look for sub-groups of index 2 in each of the 32 crystallographic point groups.

The complete list is given in Table 27.2.

There is one non-crystallographic particoloured group that should be mentioned here. The icosahedral group $\bar{5}\,\bar{3}\,\frac{2}{m}$ (of 120 members) has a subgroup 532 (of 60) and hence there is a coloured group $\bar{5}'\bar{3}'\frac{2}{m'}$. The diagrams of Fig. 20.08 (p. 129) illustrate this, the red being now interpreted as the second colour.

Space groups

It would be possible to apply the same process to the 75 crystallographic line groups, the 80 plane groups and the 230 space groups. We do not attempt this huge task but it may be of interest if we give some indication of the way in which the 230 space groups lead to the 1651 coloured groups known as the *Shubnikov groups*.

Fig. 27.08

(*a*)   (*b*)

Table 27.2

Group	Sub-group	Parti-coloured group
2	1	$2'$
4	2	$4'$
6	3	$6'$
$1m$	1	$1m'$
$2mm$	2	$2m'm'$
	$1m$	$2'mm'$
$3m$	3	$3m'$
$4mm$	4	$4m'm'$
	$2mm$	$4'mm'$
$6mm$	6	$6m'm'$
	$3m$	$6'mm'$
$2/m$	2	$2/m'$
	m	$2'/m$
	$\bar{1}$	$2'/m'$
$4/m$	4	$4/m'$
	$2/m$	$4'/m$
	$\bar{4}$	$4'/m'$
$6/m$	6	$6/m'$
	$\bar{6}$	$6'/m$
	$\bar{3}$	$6'/m'$
$\frac{2}{m}\frac{2}{m}\frac{2}{m}$	222	$\frac{2}{m'}\frac{2}{m'}\frac{2}{m'}$
	$\frac{2}{m}$	$\frac{2}{m}\frac{2'}{m'}\frac{2'}{m'}$
	$2mm$	$\frac{2}{m'}\frac{2'}{m}\frac{2'}{m}$
$\frac{4}{m}\frac{2}{m}\frac{2}{m}$	422	$\frac{4}{m'}\frac{2}{m'}\frac{2}{m'}$
	$\frac{4}{m}$	$\frac{4}{m}\frac{2'}{m'}\frac{2'}{m'}$
	$4mm$	$\frac{4}{m'}\frac{2'}{m}\frac{2'}{m}$
	$\frac{2}{m}\frac{2}{m}\frac{2}{m}$	$\frac{4'}{m}\frac{2}{m}\frac{2'}{m'}$
	$\bar{4}m2$	$\frac{4'}{m'}\frac{2'}{m}\frac{2}{m'}$

Group	Sub-group	Parti-coloured group
$\frac{2}{m}\frac{2}{m}\frac{2}{m}$	622	$\frac{6}{m'}\frac{2}{m'}\frac{2}{m'}$
	$\frac{6}{m}$	$\frac{6}{m}\frac{2'}{m'}\frac{2'}{m'}$
	$6mm$	$\frac{6}{m'}\frac{2'}{m}\frac{2'}{m}$
	$\bar{6}m2$	$\frac{6'}{m}\frac{2'}{m}\frac{2}{m'}$
	$\bar{3}\frac{2}{m}$	$\frac{6'}{m'}\frac{2}{m}\frac{2'}{m'}$
222	2	$22'2'$
32	3	$32'$
422	4	$42'2'$
	222	$4'22'$
622	6	$62'2'$
	32	$6'22'$
$\bar{1}$	1	$\bar{1}'$
$\bar{3}$	3	$\bar{3}'$
$\bar{4}$	2	$\bar{4}'$
$\bar{6}$	3	$\bar{6}'$
$\bar{3}\frac{2}{m}$	$\bar{3}$	$\bar{3}\frac{2'}{m'}$
	32	$\bar{3}'\frac{2}{m'}$
	3m	$\bar{3}'\frac{2'}{m}$
$\bar{4}m2$	$\bar{4}$	$\bar{4}m'2'$
	222	$\bar{4}'m2$
	2mm	$\bar{4}'m'2$
$\bar{6}m2$	$\bar{6}$	$\bar{6}m'2'$
	32	$\bar{6}'m2$
	3m	$\bar{6}'m'2$
$\bar{4}3m$	23	$\bar{4}'3m'$
$\frac{2}{m}\bar{3}$	23	$\frac{2}{m'}\bar{3}'$
432	23	$4'32'$
$\frac{4}{m}\bar{3}\frac{2}{m}$	432	$\frac{4}{m}\bar{3}'\frac{2}{m'}$
	$\frac{2}{m}\bar{3}$	$\frac{4'}{m}\bar{3}\frac{2'}{m'}$
	$\bar{4}3m$	$\frac{4'}{m}\bar{3}'\frac{2'}{m}$

Each of the 230 groups is based on a lattice and a point group, many of them with associated glides or screws. Each gives a single-coloured group (black or white) and a grey one, and there are usually particoloured groups obtained by colour changes either with translation or with one or more of the other movements. The possibilities for change of colour with translation are represented by 22 'black and white' lattices, in which the black points form the same pattern as the white and each point of one colour is midway between two of the opposite colour. They are shown in Fig. 27.10, in which it is understood, as before, that each point shown is repeated in every unit cell of the lattice.

To see why there are only 22 of these lattices, note first that change of colour can be along any of the edges of the cell, or the diagonals of the faces, or the 'body-diagonal'. With the triclinic lattice the edges are indistinguishable and the diagonals can become edges by different choice of the unit cell: so there is only one black-and-white lattice. With the monoclinic lattice $P2/m$ the z-axis is unique and this increases the possibilities to three. The lattice $B2/m$ is centred on the Y-face and this means that, while the x- and y-axes are now distinguishable, colour changes in the z- and x-directions cannot occur independently; and changes in the diagonal directions would produce lattices identical with the first of the two illustrated. With the orthorhombic lattices there are more possibilities, but they are limited by the same considerations. With the tetragonal lattices the x- and y-axes must be treated alike and changes in those directions would give a lattice centred on the Z-face, which reduces to a smaller tetragonal lattice with change of colour along the Z-diagonal. Similarly changes in the directions of the X- and Y-diagonals lead to a smaller lattice with change along the body-diagonal. The hexagonal and rhombohedral lattices each provide only one new type: any other changes would give black points closer together than the white ones. Finally the cubic lattices are limited to one each, for the primitive and face-centred lattices only, any other changes leading to lattices of smaller cubes, with colour changes of the types shown.

In considering colour changes within the unit cell it must be borne in mind that the symmetries are usually determined by two reflexions (or glides), or sometimes three, or by two rotations (or screws), or by one movement of each of these two kinds. These are independent movements and can carry change of colour independently of one another. Other movements are automatically determined. Triad rotation cannot contribute to particoloured groups.

To show the rich variety of coloured groups obtainable from a single uncoloured one we consider first the group $P2mm$. Here the z-direction is unique but those of x and y are indistinguishable. Change of colour can accompany one or both reflexions, the rotation being thereby determined. For change with translation we can use the y- or z-directions (P_b' or P_c') or the diagonals in the Y- or Z-planes (P_B' or P_C') or the body-diagonal (P_I'). There are thus seven particoloured groups, making nine in all, namely

$P2mm$, $P2mm1'$,
$P2'mm'$, $P2m'm'$,
$P_b'2mm$, $P_c'2mm$, $P_B'2mm$, $P_C'2mm$, $P_I'2mm$.

Similarly from the uncoloured group $P2/m$ we obtain

$P2/m$, $P2/m1'$,
$P2'/m$, $P2/m'$, $P2'/m'$,
$P_b'2/m$, $P_c'2/m$, $P_B'2/m$ $(= P_I'2/m)$.

With $P4mm$ the two sets of reflexion planes are distinguishable, so we have

$P4mm$, $P4mm1'$,
$P4'mm'$, $P4'm'm$, $P4m'm'$,
$P_c'4mm$, $P_C'4mm$, $P_I'4mm$.

Fig. 27.10

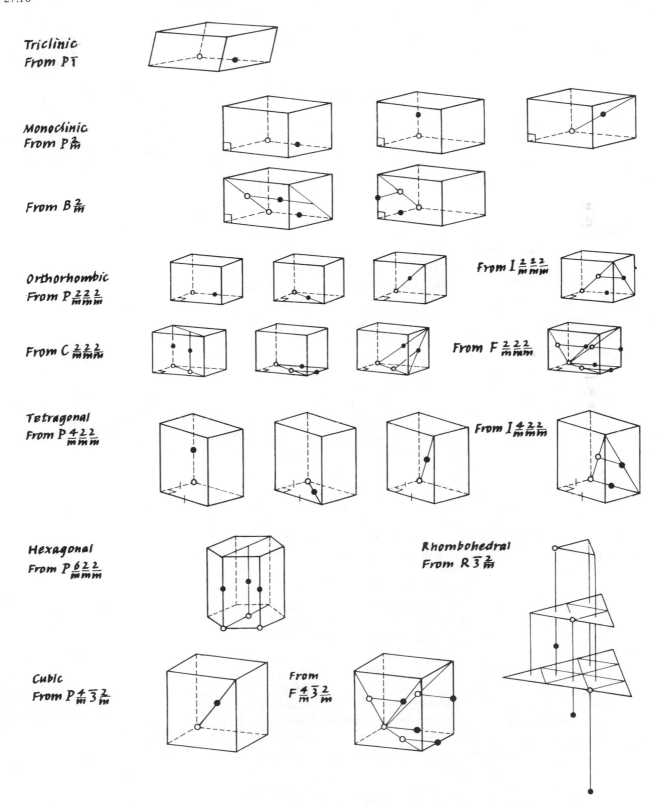

Triclinic
From $P\bar{1}$

Monoclinic
From $P\frac{2}{m}$

From $B\frac{2}{m}$

Orthorhombic
From $P\frac{2}{m}\frac{2}{m}\frac{2}{m}$

From $I\frac{2}{m}\frac{2}{m}\frac{2}{m}$

From $C\frac{2}{m}\frac{2}{m}\frac{2}{m}$

From $F\frac{2}{m}\frac{2}{m}\frac{2}{m}$

Tetragonal
From $P\frac{4}{m}\frac{2}{m}\frac{2}{m}$

From $I\frac{4}{m}\frac{2}{m}\frac{2}{m}$

Hexagonal
From $P\frac{6}{m}\frac{2}{m}\frac{2}{m}$

Rhombohedral
From $R\bar{3}\frac{2}{m}$

Cubic
From $P\frac{4}{m}\bar{3}\frac{2}{m}$

From
$F\frac{4}{m}\bar{3}\frac{2}{m}$

From $P\frac{6}{m}\frac{2}{c}\frac{2}{c}$ we derive ten coloured groups:

Fig. 27.11

$$P\frac{6}{m}\frac{2}{c}\frac{2}{c}, \qquad P\frac{6}{m}\frac{2}{c}\frac{2}{c}1',$$

$$P\frac{6'}{m}\frac{2'}{c}\frac{2}{c'}, \qquad P\frac{6'}{m}\frac{2}{c'}\frac{2'}{c}, \qquad P\frac{6}{m}\frac{2'}{c'}\frac{2'}{c}, \qquad P\frac{6}{m'}\frac{2'}{c}\frac{2'}{c}, \qquad P\frac{6'}{m'}\frac{2}{c}\frac{2'}{c'}, \qquad P\frac{6}{m'}\frac{2'}{c'}\frac{2'}{c}, \qquad P\frac{6}{m'}\frac{2}{c'}\frac{2}{c'},$$

$$P'_c\frac{6}{m}\frac{2}{c}\frac{2}{c}.$$

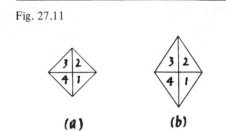

(a)   (b)

With the cubic group $P432$, however, there are only four, of which the 'grey' one can be written $P43'2$:

$$P432, \qquad P43'2,$$
$$P4'32',$$
$$P'_I432.$$

The complete list of 1651 Shubnikov groups is given in Shubnikov & Belov's *Colored Symmetry*.

Polychromatic symmetry

For polychromatic symmetry there must be, on our definition, a definite sequence of colours, and a symmetry movement of the group must either carry no change of colour or a change of one step in the sequence, or two steps, or some other fixed number of steps. This is equivalent to Belov's assumption (Shubnikov & Belov: *Colored Symmetry*, p. 228) that a coloured plane pattern can be represented three-dimensionally with the colours represented by different levels in each unit cell.

Consider, for example, the patterns shown in Fig. 27.11 (where the numbers represent colours). In that of Fig. 27.11(*a*) the colour sequence 1, 2, 3, 4 is associated with tetrad rotation; and if the colours are represented by different levels the rotation becomes a screw. The pattern of Fig. 27.11(*b*), however, does not exhibit four-colour symmetry, because it has no symmetry movement accompanied by a four-colour sequence.* If it were represented three-dimensionally, with the colours shown by four levels in the order 1, 2, 3, 4 in each unit of the colour axis, the only symmetry movement would be a diad screw, representing the coloured diad rotation. It is thus a pattern of two colour schemes rather than of four colours. (If the levels representing the colours were in the order 1, 3, 2, 4 there would be a glide reflexion in the yz-plane, representing a coloured reflexion in the y-axis in the original two-dimensional pattern.)

The method of representing three or more colours by means of an extra dimension is shown in Fig. 27.12, (*a*) for a rotation, (*b*) for a translation. In (*a*) the colour sequence black—grey—white is associated with a triad rotation in the original xy-plane. The black point A is represented three-dimensionally by a series of uncoloured points a at levels 0, 1, 2, ... Similarly the grey point B is represented by points b at levels $\frac{1}{3}$, $1\frac{1}{3}$, $2\frac{1}{3}$, . . . ; and the white point C by points c at levels $\frac{2}{3}$, $1\frac{2}{3}$, $2\frac{2}{3}$, ... These points form a three-dimensional pattern of type 3_1, so the coloured rotation is represented by an uncoloured screw. (We do not regard 3_1 and 3_2 as indicating two different types of coloured group, as it is merely a question of changing the order of the colours; but one might do so if the colours were considered to have an inherent order

---

*Koptsik, in Shubnikov & Koptsik's *Symmetry in Science and Art*, adopts a broader definition allowing, for example, two-colour changes within a four-colour pattern. He would describe the pattern of Fig. 27.11(*b*) as $(2^{(2)}m^{(2)}m^{(2)})^{(4)}$.

of their own, for instance as determined by the spectrum, or if the colours represented some other characteristics in which the order mattered.)

We designate a triad rotation in three colours as $3^{(3)}$. It is also possible to have a three-colour sequence occurring twice in a six-fold rotation, which would be $6^{(3)}$, and so on. These rotational point groups are the only polychromatic ones in two dimensions, reflexions being excluded by the polychromatic rotation. There are an infinite number of such groups, as any number of colours may be used, but when we come to repeating patterns we are confined to the crystallographic rotations 3, 4 and 6.

Polychromatic plane groups in two dimensions can be found from uncoloured space groups in three dimensions. In selecting from the 230 space groups it is to be noticed that, where rotations occur, the *z*-axis must be perpendicular to the *xy*-plane. The height of the unit cell is the interval *aa* (or *bb*, etc.) between successive points representing the same coloured point. The groups containing change of colour with rotation can therefore be obtained by selecting from the 230 space groups those with points on 3, 4 or 6 levels in each unit cell. These are eleven in number, as illustrated in Fig. 11.20 (p. 70), or 15 if the colours are considered to have their own inherent order.

Fig. 27.12

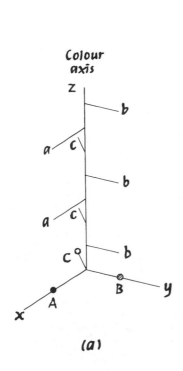

(a)

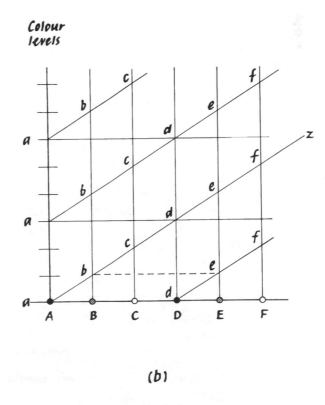

(b)

Where colour change is associated only with translation the situation is somewhat different. A three-colour change of this kind is shown in Fig. 27.12(*b*). A unit cell in this pattern has cross-section *abed*, *e* and *d* being on the same levels as *b* and *a* respectively. It thus appears that the *z*-axis is no longer perpendicular to both the others, and the unit on that axis is *ab* rather than *aa*. The space groups are found in the triclinic and monoclinic systems, with the *z*-axis oblique to the *xy*-plane. They are:

in the triclinic system $P1$,

in the monoclinic system $P1m$, $P1a$, $C1m$.

Each of these can be used with any number of colours, making four infinite sets for plane groups (wallpaper types), as illustrated in Fig. 27.13. For line groups (frieze types) $C1m$ is not available and the types of frieze pattern are represented by the top lines of the first three diagrams.

Continuous coloured groups

It remains to consider continuous and semi-continuous coloured groups. The single-coloured groups, and also the grey ones, are as already listed. We now deal only with those in which there is an actual change of colour, and this cannot accompany infinitesimal movements. Thus we shall not count the limits of $r'2$ and n' (Fig. 27.14), which we regard as being of the grey type; but $r2'$ and $r11m'$ provide a limiting group, of the type shown in Fig. 27.15.

In two dimensions there are, on this basis, no continuous point groups; and in a frieze group colour change is possible with reflexion in a longitudinal, but not with a transverse, mirror line. There is in fact just one continuous coloured frieze group, as illustrated in Fig. 27.15.

In a wallpaper group, too, a dichromatic change may accompany reflexion in a mirror line in the direction of the continuous translation (Fig. 27.16); or it may be linked with the discontinuous translation, as in the two types shown in Fig. 27.17. There are also groups of this latter kind using a sequence of three or more colours.

Fig. 27.15

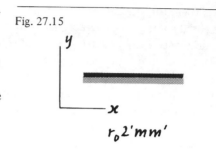

$r_0 2'mm'$

Fig. 27.16

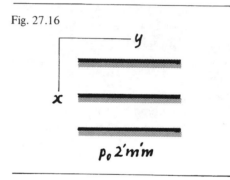

$p_0 2'm'm$

Fig. 27.14

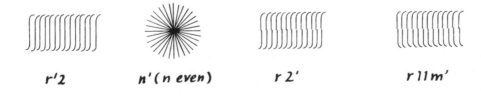

$r'2$ $n'(n\ even)$ $r2'$ $r11m'$

Fig. 27.17

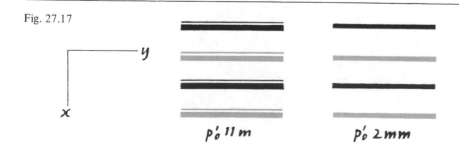

$p'_0 11m$ $p'_0 2mm$

Fig. 27.13

Triclinic
From P1

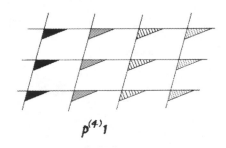

$p^{(4)}1$

Monoclinic
From P1m

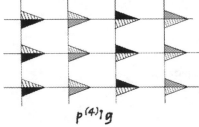

$p^{(4)}1m$

From P1a

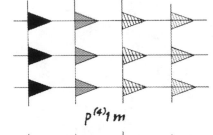

$p^{(4)}1g$ $p^{(3)}1g$

From C1m

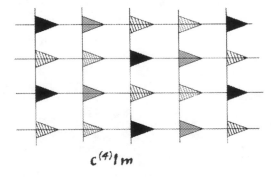

$c^{(4)}1m$ $c^{(3)}1m$

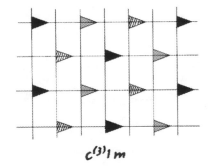

In three dimensions there are two uncoloured continuous point groups and the one that includes reflexion in the z-plane leads to a dichromatic group (Fig. 27.18).

For a rod group the same point group can be combined with a discontinuous translation, giving $r\frac{\infty}{m}\frac{2'}{m}$. Alternatively, if the discontinuous translation is linked with the change of colour, we may use either of the single-coloured continuous point groups. The dichromatic groups of this kind are $r'\infty m$ and $r'\frac{\infty}{m}\frac{2}{m}$ (Fig. 27.19), but any number of colours may be used.

With continuous translation the cross-section of the pattern is a two-dimensional 'rosette', based on any point group of the series n or nm (n odd) or nmm (n even). Change of colour may be associated with the rotation, the number of colours being n or a factor of n. The dichromatic groups (Fig. 27.20) are $r_0\frac{n'}{m}$ and $r_0\frac{n'}{m}\frac{22'}{mm}$ (n even). The screw types $r\infty_n^+2$ and $r\infty_n^-2$ may be coloured in the same way.

For continuous layer patterns there is, for our purposes, only one dichromatic type, with cross-section as shown in Fig. 27.15. For the semi-continuous patterns there is the same range of dichromatic types as for friezes, and in addition an infinite number of types in which changes of two or more colours are associated with the discontinuous translations.

Last of all there are the coloured three-dimensional space groups. With continuity in one direction the possibilities are the same as for the wallpaper groups. If it is in two directions, and hence in all directions in a plane, the cross-section may be dichromatic, as in Fig. 27.16, or there may be a change of two or more colours with the discontinuous translation, as in Fig. 27.17. But if there is continuity in all directions in space there cannot be any colour changes at all.

Fig. 27.18

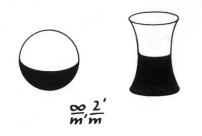

$$\frac{\infty}{m},\frac{2'}{m}$$

Fig. 27.19

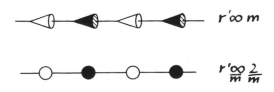

$r'\infty m$

$r'\frac{\infty}{m}\frac{2}{m}$

Fig. 27.20

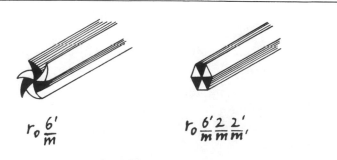

$r_0\frac{6'}{m}$ $r_0\frac{6'2}{m\,m}\frac{2'}{m'}$

Appendix 1

The symmetry group of a finite pattern is a point group

In two dimensions

The possible movements are

      rotation, reflexion,

      translation, glide reflexion.

Of these, the last two cannot bring a finite pattern into self-coincidence. So, apart from the identity, the group can contain only rotations and reflexions.

    Rotations can be about one centre only:

for if S, S' were rotations through angles 2α, 2β about A, B respectively (Fig. A.1), SS' would be a rotation of amount $2\alpha + 2\beta$ about C (in the figure) and $S'S$ an equal one about C'. The combination of SS' and $(S'S)^{-1}$ would be a translation. But the group cannot contain a translation.

    If there is a rotation centre, any reflexion line must pass through it:

for otherwise the reflexion would imply a second rotation centre.

    If there are no rotations, there can be only one reflexion line:

for two would imply either a translation or a rotation.

    The conclusion is that the group may either consist of the identity and a single reflexion, or it may contain rotations about one centre, with or without reflexions in lines through that centre. It is thus necessarily a point group.

In three dimensions

The possible movements are:

      rotation, reflexion, inversion, rotatory inversion,

      translation, glide reflexion, screw rotation.

Of these, the last three cannot bring a finite pattern into self-coincidence. So, apart from the identity, the group can contain movements of the first four types only.

    Any two rotation axes meet in a point:

for if not, they would be either parallel or skew.

    (i) Suppose the axes were parallel. The argument based on Fig. A.1 would apply and the group would contain a translation.

    (ii) Suppose the axes were skew. The combined movement would be a direct movement and hence either a rotation, a translation or a screw. It is sufficient to show that it could not be a rotation. A point at P on the axis of such a resultant rotation would go to P' (say) under the first movement and back to P under the second. Either P' would coincide with P, in which case both the original rotation axes would pass through that point, contrary to hypothesis; or, if not, both axes would be in the plane bisecting PP' at right angles, again contrary to hypothesis.

Fig. A.1

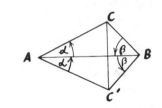

Any further rotation axis must pass through the same point:
for otherwise it would be skew to S or S' or SS'.

If in addition to two or more rotation axes there are reflexion planes,
they must pass through the point of intersection of the axes:
for otherwise there would be further rotation axes not through that point.

If there is only one rotation axis, any reflexion plane must either be
perpendicular to it or contain it:
for otherwise there would be another rotation axis.

There can be only one such plane perpendicular to that axis:
for parallel reflexion planes imply a translation.

If there are no rotations there can be only one reflexion plane:
for two imply either a translation or a rotation.

If there is an inversion or a rotatory inversion, the centre of inversion
must lie on any rotation axis and in any reflexion plane:
for otherwise there would be further axes or reflexion planes not satisfying the
above conditions.

The conclusion is that, in the symmetry group of a finite pattern, there is
always at least one point that is left unchanged by any of the movements of the
group. The group is therefore a point group.

Appendix 2

Matrices

Using coordinates, a transformation is best expressed by a matrix. If the point at the origin remains fixed a 3×3 matrix can be used, constructed as follows. (We here use coordinates in the order x, y, z.)

> In the first column put the coordinates of the point to which $(1, 0, 0)$ is moved; in the second, those for the point $(0, 1, 0)$; and in the third, those for $(0, 0, 1)$.

Thus for a rotation of $+90°$ about the axis of y we have

$$\begin{pmatrix} 0 & 0 & 1 \\ 0 & 1 & 0 \\ -1 & 0 & 0 \end{pmatrix}.$$

Applying this to the point (x, y, z),

$$\begin{pmatrix} 0 & 0 & 1 \\ 0 & 1 & 0 \\ -1 & 0 & 0 \end{pmatrix} \begin{pmatrix} x \\ y \\ z \end{pmatrix} = \begin{pmatrix} z \\ y \\ -x \end{pmatrix}.$$

For the same rotation followed by inversion in the origin we may multiply by the inversion matrix:

$$\begin{pmatrix} -1 & 0 & 0 \\ 0 & -1 & 0 \\ 0 & 0 & -1 \end{pmatrix} \times \begin{pmatrix} 0 & 0 & 1 \\ 0 & 1 & 0 \\ -1 & 0 & 0 \end{pmatrix} = \begin{pmatrix} 0 & 0 & -1 \\ 0 & -1 & 0 \\ 1 & 0 & 0 \end{pmatrix}.$$

Applying this to (x, y, z),

$$\begin{pmatrix} 0 & 0 & -1 \\ 0 & -1 & 0 \\ 1 & 0 & 0 \end{pmatrix} \begin{pmatrix} x \\ y \\ z \end{pmatrix} = \begin{pmatrix} -z \\ -y \\ x \end{pmatrix}.$$

(This is of course the previous result with all three signs changed.)

For point groups we need only the following matrices:

With rectangular axes

Rotations about Oz		*Diad rotations*	
Diad	*Tetrad*	*about Oy*	*about $x=y$, $z=0$*

$$\begin{pmatrix} -1 & 0 & 0 \\ 0 & -1 & 0 \\ 0 & 0 & 1 \end{pmatrix} \quad \begin{pmatrix} 0 & -1 & 0 \\ 1 & 0 & 0 \\ 0 & 0 & 1 \end{pmatrix} \quad \begin{pmatrix} -1 & 0 & 0 \\ 0 & 1 & 0 \\ 0 & 0 & -1 \end{pmatrix} \quad \begin{pmatrix} 0 & 1 & 0 \\ 1 & 0 & 0 \\ 0 & 0 & -1 \end{pmatrix}$$

Triad rotation   *Inversion*
about $x=y=z$

$$\begin{pmatrix} 0 & 0 & 1 \\ 1 & 0 & 0 \\ 0 & 1 & 0 \end{pmatrix} \quad \begin{pmatrix} -1 & 0 & 0 \\ 0 & -1 & 0 \\ 0 & 0 & -1 \end{pmatrix}$$

With axes as for hexagonal lattices

Rotations about Oz   *Diad rotation*   *Inversion*

Triad   *Hexad*   *about Oy*

$$\begin{pmatrix} 0 & -1 & 0 \\ 1 & -1 & 0 \\ 0 & 0 & 1 \end{pmatrix} \quad \begin{pmatrix} 1 & -1 & 0 \\ 1 & 0 & 0 \\ 0 & 0 & 1 \end{pmatrix} \quad \begin{pmatrix} -1 & 0 & 0 \\ -1 & 1 & 0 \\ 0 & 0 & -1 \end{pmatrix} \quad \begin{pmatrix} -1 & 0 & 0 \\ 0 & -1 & 0 \\ 0 & 0 & -1 \end{pmatrix}$$

Using these matrices we can write down the coordinates of sets of equivalent points in any of the groups:

Point Group

1	*2*	*4*	*3*	*6*
x, y, z	x, y, z	x, y, z	x, y, z	x, y, z
	$-x, -y, z$	$-x, -y, z$	$-y, x-y, z$	$-y, x-y, z$
		$-y, x, z$	$y-x, -x, z$	$y-x, -x, z$
		$y, -x, z$		$x-y, x, z$
				$-x, -y, z$
				$y, y-x, z$

222	*422*	*32*	*622*	
x, y, z	x, y, z	x, y, z	x, y, z	$-x, y-x, -z$
$-x, -y, z$	$-x, -y, z$	$-y, x-y, z$	$-y, x-y, z$	$y, x, -z$
$-x, y, -z$	$-y, x, z$	$y-x, -x, z$	$y-x, -x, z$	$x-y, -y, -z$
$x, -y, -z$	$y, -x, z$	$-x, y-x, -z$	$x-y, x, z$	$y-x, y, -z$
	$-x, y, -z$	$y, x, -z$	$-x, -y, z$	$x, x-y, -z$
	$x, -y, -z$	$x-y, -y, -z$	$y, y-x, z$	$-y, -x, -z$
	$y, x, -z$			
	$-y, -x, -z$			

23. As for 222, with cyclic permutations of the letters (12 points in all).

432. As for 422, with cyclic permutations (24 points in all).

For the next eleven groups (Fig. 20.07, p. 128) we add inversion, changing the sign of all three coordinates of each point in a group. This doubles the number of points.

For the remaining ten groups (Fig. 20.10, p. 130) we follow the method described on p. 129, replacing half the points in a group by their inversions. Thus, for example, to obtain the points for *4mm*, we take the first four of those given for 422, and change all signs in the remaining four. The result is:

x, y, z	$x, -y, z$
$-x, -y, z$	$-x, y, z$
$-y, x, z$	$-y, -x, z$
$y, -x, z$	y, x, z

If the point at the origin is moved to (h, k, l) we use a 4×4 matrix constructed as follows.

We write down the 3×3 matrix for the corresponding movement with the point at the origin fixed, and add a border, the fourth elements of the first three rows being h, k, l, and the fourth row 0, 0, 0, 1.

Thus, for example, a diad screw about the axis $x = \frac{1}{4}$, $y = 0$ (see Fig. 25.19) (p. 182) is represented by the matrix

$$\begin{pmatrix} -1 & 0 & 0 & \frac{1}{2} \\ 0 & -1 & 0 & 0 \\ 0 & 0 & 1 & \frac{1}{2} \\ 0 & 0 & 0 & 1 \end{pmatrix}$$

If this movement is followed by a diad screw about $x = \frac{1}{2}$, $z = \frac{1}{4}$, as in the same figure, we have

$$\begin{pmatrix} -1 & 0 & 0 & 1 \\ 0 & 1 & 0 & \frac{1}{2} \\ 0 & 0 & -1 & \frac{1}{2} \\ 0 & 0 & 0 & 1 \end{pmatrix} \times \begin{pmatrix} -1 & 0 & 0 & \frac{1}{2} \\ 0 & -1 & 0 & 0 \\ 0 & 0 & 1 & \frac{1}{2} \\ 0 & 0 & 0 & 1 \end{pmatrix} = \begin{pmatrix} 1 & 0 & 0 & \frac{1}{2} \\ 0 & -1 & 0 & \frac{1}{2} \\ 0 & 0 & -1 & 0 \\ 0 & 0 & 0 & 1 \end{pmatrix},$$

showing that the combined movement is another diad screw, about an axis $y = \frac{1}{4}$, $z = 0$.

As a second example we consider tetrad screws about the axes shown in Fig. 25.24(*a*) (p. 184). If S_1 is a 4_1 screw about the axis $z = 0$, $y = \frac{1}{4}$, and S_2 another about $y = 0$, $x = \frac{1}{4}$, the product $S_2 S_1$ is given by

$$\begin{array}{ccc} S_2 & S_1 & S_2 S_1 \\ \begin{pmatrix} 0 & -1 & 0 & \frac{1}{4} \\ 1 & 0 & 0 & -\frac{1}{4} \\ 0 & 0 & 1 & \frac{1}{4} \\ 0 & 0 & 0 & 1 \end{pmatrix} \times & \begin{pmatrix} 1 & 0 & 0 & \frac{1}{4} \\ 0 & 0 & -1 & \frac{1}{4} \\ 0 & 1 & 0 & -\frac{1}{4} \\ 0 & 0 & 0 & 1 \end{pmatrix} = & \begin{pmatrix} 0 & 0 & 1 & 0 \\ 1 & 0 & 0 & 0 \\ 0 & 1 & 0 & 0 \\ 0 & 0 & 0 & 1 \end{pmatrix}. \end{array}$$

representing a triad rotation about $x = y = z$. The transform of S_1 by S_2, namely $S_2 S_1 S_2^{-1}$, is then found by

$$\begin{array}{ccc} S_2 S_1 & S_2^{-1} & S_2 S_1 S_2^{-1} \\ \begin{pmatrix} 0 & 0 & 1 & 0 \\ 1 & 0 & 0 & 0 \\ 0 & 1 & 0 & 0 \\ 0 & 0 & 0 & 1 \end{pmatrix} \times & \begin{pmatrix} 0 & 1 & 0 & \frac{1}{4} \\ -1 & 0 & 0 & \frac{1}{4} \\ 0 & 0 & 1 & -\frac{1}{4} \\ 0 & 0 & 0 & 1 \end{pmatrix} = & \begin{pmatrix} 0 & 0 & 1 & -\frac{1}{4} \\ 0 & 1 & 0 & \frac{1}{4} \\ -1 & 0 & 0 & \frac{1}{4} \\ 0 & 0 & 0 & 1 \end{pmatrix}. \end{array}$$

which is a 4_1 screw about $x = 0$, $z = \frac{1}{4}$.

The cubic groups are completely determined by the tetrad rotations or rotatory inversions. For example, in $P\bar{4}3c$ (Fig. 25.30, p. 187) we note first that a $\bar{4}$ movement about the y-axis, with inversion centre at the origin, would have the matrix

$$\begin{pmatrix} 0 & 0 & -1 \\ 0 & -1 & 0 \\ 1 & 0 & 0 \end{pmatrix}.$$

For the same movement about the parallel axis $x = \frac{1}{2}$, $z = 0$, with inversion centre at $(\frac{1}{2}, \frac{1}{4}, 0)$, the bordered matrix is

$$\begin{pmatrix} 0 & 0 & -1 & \frac{1}{2} \\ 0 & -1 & 0 & \frac{1}{2} \\ 1 & 0 & 0 & -\frac{1}{2} \\ 0 & 0 & 0 & 1 \end{pmatrix}.$$

If this movement is followed by a turn of $+90°$ about $y = \frac{1}{2}$, $z = 0$, with inversion in $(\frac{1}{4}, \frac{1}{2}, 0)$, the resultant movement is obtained by the multiplication

$$
\begin{pmatrix} -1 & 0 & 0 & \frac{1}{2} \\ 0 & 0 & 1 & \frac{1}{2} \\ 0 & -1 & 0 & \frac{1}{2} \\ 0 & 0 & 0 & 1 \end{pmatrix} \times \begin{pmatrix} 0 & 0 & -1 & \frac{1}{2} \\ 0 & -1 & 0 & \frac{1}{2} \\ 1 & 0 & 0 & -\frac{1}{2} \\ 0 & 0 & 0 & 1 \end{pmatrix} = \begin{pmatrix} 0 & 0 & 1 & 0 \\ 1 & 0 & 0 & 0 \\ 0 & 1 & 0 & 0 \\ 0 & 0 & 0 & 1 \end{pmatrix},
$$

giving a triad rotation as before. Multiplying again by the original matrix we obtain

$$
\begin{pmatrix} 0 & -1 & 0 & \frac{1}{2} \\ -1 & 0 & 0 & \frac{1}{2} \\ 0 & 0 & 1 & -\frac{1}{2} \\ 0 & 0 & 0 & 1 \end{pmatrix}, \text{ which changes } \begin{pmatrix} x \\ y \\ z \\ 1 \end{pmatrix} \text{ to } \begin{pmatrix} \frac{1}{2}-y \\ \frac{1}{2}-x \\ -\frac{1}{2}+z \\ 1 \end{pmatrix}
$$

and thus represents a c-glide in the plane $x + y = \frac{1}{2}$.

For the group $I\bar{4}3d$ we rotate $+90°$ about $y = \frac{1}{4}$, $z = 0$ and invert in $(\frac{1}{8}, \frac{1}{4}, 0)$; then rotate $-90°$ about $z = \frac{1}{4}$, $x = 0$ and invert in $(0, \frac{1}{8}, \frac{1}{4})$.

$$
\begin{pmatrix} 0 & 0 & 1 & -\frac{1}{4} \\ 0 & -1 & 0 & \frac{1}{4} \\ -1 & 0 & 0 & \frac{1}{4} \\ 0 & 0 & 0 & 1 \end{pmatrix} \times \begin{pmatrix} -1 & 0 & 0 & \frac{1}{4} \\ 0 & 0 & 1 & \frac{1}{4} \\ 0 & -1 & 0 & \frac{1}{4} \\ 0 & 0 & 0 & 1 \end{pmatrix} = \begin{pmatrix} 0 & -1 & 0 & 0 \\ 0 & 0 & -1 & 0 \\ 1 & 0 & 0 & 0 \\ 0 & 0 & 0 & 1 \end{pmatrix},
$$

showing that the product movement is a triad rotation about the line $x = -y = z$. Multiplying again by the second matrix,

$$
\begin{pmatrix} -1 & 0 & 0 & \frac{1}{4} \\ 0 & 0 & 1 & \frac{1}{4} \\ 0 & -1 & 0 & \frac{1}{4} \\ 0 & 0 & 0 & 1 \end{pmatrix} \times \begin{pmatrix} 0 & -1 & 0 & 0 \\ 0 & 0 & -1 & 0 \\ 1 & 0 & 0 & 0 \\ 0 & 0 & 0 & 1 \end{pmatrix} = \begin{pmatrix} 0 & 1 & 0 & \frac{1}{4} \\ 1 & 0 & 0 & \frac{1}{4} \\ 0 & 0 & 1 & \frac{1}{4} \\ 0 & 0 & 0 & 1 \end{pmatrix}.
$$

This changes (x, y, z) to $(y+\frac{1}{4}, x+\frac{1}{4}, z+\frac{1}{4})$, representing a diamond glide and, by repetition, an I-lattice.

Books

H. Weyl: *Symmetry* (Princeton, 1952).
  An attractive introduction.
J. Rosen: *Symmetry Discovered* (Cambridge, 1975).
  A general, rather than detailed, treatment.
F.C. Phillips: *Introduction to Crystallography* (Oliver & Boyd, 1971).
  A full descriptive account of crystal forms.
A. Holden: *Shapes, Space and Symmetry* (Columbia, New York, 1971).
  Deals with regular and other solids, the filling of space, lattices, knots, etc. Well illustrated.
M. Kraitchik: *Mathematical Recreations* (Allen & Unwin, 1960).
  Contains an excellent chapter on mosaics.
A. Bell & T. Fletcher: *Symmetry Groups* (Assoc. of Teachers of Mathematics, 1970).
  A brief elementary introduction, from the mathematical point of view.
F.J. Budden: *The Fascination of Groups* (Cambridge, 1972).
  An introduction to groups, with chapters on symmetry and patterns.
A.V. Shubnikov & N.V. Belov: *Colored Symmetry* (Pergamon, 1964, 1968).
  Translations from the Russian of original work on coloured symmetry; with extensive
  bibliography up to 1968.
A.V. Shubnikov & V.A. Koptsik: *Symmetry in Science & Art* (Plenum Press, New York,
    1974).
  A comprehensive work covering all kinds of symmetry groups, with a considerable section
  on group theory. About 500 references, up to 1974.
International Tables for X-ray Crystallography, vol. 1 (Kynoch Press, Birmingham, 3rd edn
    1969).
  The standard reference book.
W.L. Bragg: *An Introduction to Crystal Analysis* (Bell, London, 1928).
  A classic. Still the best introduction.
Owen Jones: *The Grammar of Ornament* (1856. Reprinted Van Nostrand, New York, 1972).
  A classic in its field, handsomely produced and illustrated.
J. Bourgoin: *Arabic Geometrical Pattern & Design* (Firmin-Didot, Paris, 1879, Dover, New
    York, 1973).
  190 pages of line drawings of mosaics.
David Wade: *Pattern in Islamic Art* (Studio Vista, London, 1976).
  Contains many coloured and other illustrations.
D. Hill & O. Graber: *Islamic Architecture and its Decorations* (Faber, 1967).
  A well-illustrated survey.
J.L. Locher: *The World of M.C. Escher* (H.N. Abrams, New York, 1971).
  A general illustrated account of Escher's work.
C.H. Macgillavry: *Symmetry Aspects of M.C. Escher's Periodic Drawings* (Bohn, Scheltema
    & Holkema, Utrecht, 1976).
  Essential reading for the student of symmetry.
d'Arcy Thompson: *On Growth & Form* (Cambridge, Abr. Edn, 1961).
  Mathematical aspects in biology, with a wealth of examples and illustrations. A classic.
Vicenzo de Michele: *Crystals: Symmetry in the Mineral Kingdom* (Orbis, London, 1972).
  Beautiful illustrations of crystals.

Summary tables

Direct and opposite movements

Dimensions	Direct	Opposite
2	Rotation Translation Identity	Reflexion Glide reflexion
3	Rotation Translation Screw rotation Identity	Reflexion Glide reflexion Inversion Rotatory inversion

Symmetry movements without translation

Dimensions	(1) Reflexion in	(2) Rotation about	(3) Inversion in
1	a point		
2	a line	a point	
3	a plane	a line	a point

The numbers in parentheses are the numbers of coordinates whose signs are changed to effect the movement.

Symmetry movements with translation

Dimensions	Translation along a line combined with reflexion	rotation
2	Glide reflexion in the line	
3	Glide reflexion in a plane containing the line	Screw rotation about the line

Classification of symmetry groups, with numbers in each class

		Number of independent translation directions			
Dimensions	Mode of repetition	0 Point groups	1 Line groups	2 Plane groups	3 Space groups
1	1(1) (row)	2(1)	2(3)		
2	5(5) (nets)	†10(11)	7(17) (friezes)	17(46) (wallpapers)	
3	14(22) (lattices)	†32(58) (crystal forms)	†75 (rods)	80 (layers)	230(1191) (crystal structures)

†Crystallographic groups only.

In each class the first figure gives the number of uncoloured groups and the figure in parentheses the number of particoloured groups. The total number of dichromatic groups is thus found in each case by adding the second figure to twice the first. For example, the number of coloured friezes is $2 \times 7 + 17 = 31$, which is equal to the number of uncoloured two-sided bands; and it has also been found that the number of coloured bands is $2 \times 31 + 117 = 179$.

The number of coloured line groups in one dimension (7) is equal to the number of uncoloured friezes and the number of coloured plane groups in two dimensions (80) is equal to the number of uncoloured layers. Similarly the number of coloured space groups in three dimensions (1651) must be equal to the number of uncoloured three-dimensional groups in four dimensions.

Notation and axes

In the symbols there may be first a prefix, to indicate the row, net or lattice. There are then three positions, each referring to an axis of rotation, or a set of such axes, and mirror lines or mirror planes perpendicular to them. When there are both rotation and reflexion a fraction-like form is used.

Two dimensions

Axes x, y.

Point groups (finite patterns)

No prefix.

1st position: a numeral for rotation about the origin.

2nd position: m if there is reflexion in a line perpendicular to the x-axis and lines related to it by the rotation, or 1 if there is not.

3rd position: m for reflexion in lines bisecting the angles formed by those just mentioned.

i.e. if the rotation is 2, 2nd position is for reflexion in y-axis, 3rd in x-axis;
   if the rotation is 4, 2nd is for both axes, 3rd for bisectors;
   if rotation is 3 or 6, axes are at $120°$, with an extra one (u-axis) at $120°$ to
   each. 2nd position is for mirror lines perpendicular to the three axes, 3rd
   for the axes themselves as mirror lines.

Line groups (frieze patterns)

Prefix: r (row).

1st position: 2 if there is diad rotation, or 1 if not.

2nd position: m if there is reflexion in transverse mirror lines, or 1 if there is not.

3rd position: m for reflexion, or g for glide reflexion, in a longitudinal mirror line.

Plane groups (wallpaper patterns)

Prefix: p (primitive net) or c (centred net).

Three positions as for point groups.

For reflexions and glide reflexions in parallel mirror lines, m is used, rather than g.

Three dimensions

Axes z, x, y (in that order), the z-axis being an axis of smallest rotation.

Planes Z, X, Y perpendicular to the axes of z, x, y respectively.

Point groups (finite patterns)

No prefix.

1st position: a numeral for rotation about the z-axis and m if there is reflexion in the Z-plane (but for the cubic groups this position is for all three co-ordinate axes and planes perpendicular to them).

2nd position: a numeral for rotation about the x-axis and lines related to it by the z-rotation, and m if there is reflexion in planes perpendicular to those lines (but for the cubic groups 3 or $\bar{3}$ for triad rotation about the lines $z = \pm x = \pm y$).

3rd position: a numeral for rotation about lines bisecting angles formed by those of the 2nd position, and m if there is reflexion in planes perpendicular to those lines (but for the cubic groups the lines $x = y, z = 0$, etc.).

Line groups (rod patterns)

Prefix r.

z-axis in the direction of the translations.

Three positions as before, with g for glide reflexions and 4_1 etc. for screw rotations (see below).

Plane groups (layer patterns)

Prefix p or c.

z-axis perpendicular to the plane.

Three positions as before, with a or b for glides in the x- or y-directions, or g for both, or n for a diagonal direction.

Space groups (crystal structures)

Prefix P (primitive), or C, A, B (centred in xy-, yz-, zx- planes respectively), or F (centred in all three planes), or I (body-centred).

Three positions as before, with c, a, b for glides in the z-, x-, y-directions respectively, n for diagonal glides and d for diamond glides.

Screws

n_r signifies a rotation of $2\pi/n$ combined with an axial movement of r/n units.
For rotations and screws about parallel axes the rotation symbol is used.

Rotatory inversion

$\bar{n}$ signifies rotation through $2\pi/n$ combined with inversion. $\bar{1}$ is pure inversion; $\bar{2}$ is reflexion in a plane perpendicular to the axis of rotation.

Coloured groups

Dichromatic

A prime applied to the symbol for any movement means that the movement is accompanied by a change of colour. A prime applied to the prefix signifies change of colour with translation. (See Figs. 27.07, 27.09, pp. 199, 200.)

Polychromatic

Instead of a prime the number of colours is put in parentheses as a superscript. (See Figs. 11.20, 27.13, pp. 70, 209.)

Index of groups, nets and lattices

Numerals in roman type refer to pages, those in italics to diagrams. An asterisk indicates a list, with or without diagrams. For the uncoloured point groups the corresponding 'group' notation is shown in brackets.

TWO DIMENSIONS

Point groups (illustrated by finite patterns)
Uncoloured, 11–12, 106–7

	15.01*, 15.03*	
2	*Title page, 2.02*	$[C_2]$
3	*1.04*	$[C_3]$
4	*2.01, 12.16(8)*	$[C_4]$
8	*12.16 (3)*	$[C_8]$
1m	*vi, 2.03, 12.16(2)*	$[D_1$ or $C_2]$
2mm	*2.07, 12.16(1)*	$[D_2]$
3m	*2.07*	$[D_3]$
4mm	*2.07*	$[D_4]$
5m	*2.06*	$[D_5]$
6mm	*12.16 (4)*	$[D_6]$

continuous, 53, 191

∞m	*9.02*	$[D_\infty]$

Dichromatic, 59–61, 61*, 195, 196–7, 198*

	11.06
4′	*27.01(a)*
6′	*12.16 (5) (10)*
2mm1′	*11.03(c)*
2m′m′	*11.01(b) (c) (d)*
2′m′m	*12.16 (7)*
4′m′m	*12.16 (9)*
6m′m′	*12.16 (6)*

Polychromatic, 69, 195, 206–7

$4^{(4)}$	*11.19(a)*
$6^{(3)}$	*27.01(b)*
$8^{(4)}$	*11.19(b)*

Line groups (frieze patterns)
Uncoloured, 13–15, 108–9

	16.05*
r1	*3.01, 3.08*
r2	*3.08, 12.16 (14)*
r1m	*3.08, 12.16 (13)*
r11m	*3.08, 12.16 (11)*
r11g	*3.08*
r2mm	*3.08*
r2mg	*3.08, 12.11, 12.16 (12) (15)*

continuous, 53, 191

$r_0 1m$	*9.03*
$r_0 2mm$	*9.03*

THREE DIMENSIONS

Point groups (finite patterns)

General index